Julia for Machine Learning

Zacharias Voulgaris

Technics Publications

Published by:

TECHNICS PUBLICATIONS

TECHNOLOGY / LEADERSHIP

2 Lindsley Road, Basking Ridge, NJ 07920 USA

https://www.TechnicsPub.com

Edited by Jessica McCurdy-Crooks

Cover design by Lorena Molinari

First Printing 2020

ISBN, print ed.	9781634628136
ISBN, Kindle ed.	9781634628143
ISBN, ePub ed.	9781634628150
ISBN, PDF ed.	9781634628167

Library of Congress Control Number: 2020935538

Dedicated to all those who have been contributing to the Julia language and all those who have been using it for (data) science pursuits.

Contents at a Glance

Contents

Introduction

When I wrote the book *Julia for Data Science* a few years ago, it was an innovative book where few people believed in this language enough to write a whole book on the subject. Also, existing books at that time focused on Julia as a programming language rather than as a tool for data science work. Fortunately, things have changed since then, especially as the language matured, and its merits became more known.

Nowadays there are plenty of books on Julia, and lately, even courses offered by a branch of Julia Computing called Julia Academy (www.juliaacademy.com). What's more, there are courses on Julia in various universities around the world. Recently, there was a book on Julia and its application to Statistics.

After attending the Julia conference in the summer of 2018 in London, where Julia 1.0 made its debut, I started considering writing another book on the language. I was thinking of a second edition to *Julia for Data Science*, but after looking into the latest trends and the vastness of Machine Learning, I decided it would be best to focus on this particular aspect of data science. After all, with the popularity of Artificial Intelligence (AI) methods in Machine Learning (ML), it seems that data engineering is not as important as it used to be, while there is so much material on ML methods, that if I were to write a second edition of the Julia for Data Science book, it would be too large!

Anyway, parallel to the books, videos, and other materials I developed since then, including an entry in Springer's online encyclopedia on Big Data, on the Julia language (https://bit.ly/2VkzjZJ), I did some (non-academic) research in data science. Naturally, I used Julia for this, as well as many proof-of-concept projects I carried out during that time. So, when I write about Julia, I have a very practical approach to the language. What's more, I'm a practicing data scientist, not a developer or a professor, so my take on the language is very hands-on and always related to data science from a holistic perspective.

There are many books that talk about the programming aspects of Julia and some of them do a good job at it, but few books talk about how it can be useful to a data scientist or a Machine Learning professional—now enter this book. This book will be a useful aid to data science professionals who wish to use Julia in their work, be it data science or any data analytics task in general.

Since you are reading this book, you must be at least a bit curious about the Julia programming language and its usefulness in Machine Learning. But chances are that you need to be convinced of Julia's power. Everyone talks about the merits of Python, Scala, R, and even MATLAB, when it comes to data science work— where does Julia fit?

In addition, Julia has fewer packages than the above languages, because it is very easy to code from scratch. Perhaps, that's why it appeals so much to developers, particularly those who are more creative and curious to explore. Not only is it easy to code, but the code is very fast, both as a prototyping language and for executing existing code. In Python or R, for example, custom scripts are not encouraged, especially when it comes to loops, due to slow performance. Besides, most of the packages in these languages are developed in some low-level language (usually C or C++), making the corresponding scripts lightning fast. In Julia, most of the packages are written in Julia itself since you don't need to rely on a low-level language for a performance boost.

In addition, Julia empowers its users to do their own thing. Not just building a Machine Learning model from scratch, which we'll do later on in this book, but also developing new algorithms, new heuristics, and even new data science systems leveraging Machine Learning algorithms to use in your work. Without the performance boost of Julia and the ease of use of its coding syntax and grammar, this wouldn't have been as feasible a task. Nevertheless, if you want to rely solely on existing packages for your work, you have this option too, as we'll see in a later part of this book.

Julia is also quite mature when it comes to scientific work, including data science. Even before its production-ready release (ver. 1.0), it was used in scientific research and had managed to become one of the languages that the MXNet framework supported with APIs (https://bit.ly/3c4yewh). Soon after that, there was even an API for Spark, making Julia a relevant language for different kinds of data science work (https://bit.ly/2zUw8k9). Now, things have escalated since the language has plenty more packages for processes related to data science, including many Machine Learning models.

Julia is also quite easy to learn. Although many programming languages require a certain familiarity with programming, Julia is fairly straightforward right off the shelf. If someone has coded in Python, R, or MATLAB, it will seem even easier to learn, since it shares a lot of the syntax, particularly with MATLAB. Besides, many of the high-level languages used in data science have a similar structure, something that Julia shares to a certain extent, even if it follows a somewhat different programming style, namely the functional paradigm.

Compatibility with other programming languages is another advantage that Julia has as a language. Julia scripts can be called from other languages, like Python and R, while you can call scripts from other languages (including C), via the Julia kernel, making the integration of Julia with existing pipelines feasible and practical, and greatly facilitating your learning of Julia since you don't have to jump to this new programming ecosystem and abandon everything you've developed in other languages already. Appendix B covers how other languages interact with Julia.

One of the key differentiators of Julia from other data science languages is *multiple dispatch*. This is a very useful feature of the language enabling the programmer to have functions with the same name but using different inputs. The idea is that the user of these functions doesn't have to remember numerous names for specialized versions of the same function, as Julia itself decides which one to use, based on the inputs provided. Multiple dispatches, therefore, allow for cleaner code and easier maintenance of scripts, enabling the user to leverage the functional programming paradigm, saving time and effort.

Julia's built-in package manager is another reason why this programming language is a good choice when it comes to data science work. Since the use of packages is commonplace in all high-level languages, Julia was designed with this in mind. As a result, it is possible to add, remove, and update packages of the language within its kernel, using simple commands. If you use a lot of external code for your work, this feature is particularly useful, especially today, when a lot of the code that used to be in the base package of Julia is found in external packages instead. Also, the current package manager module is much faster than it used to be, so updating existing packages is a walk in the park. Naturally, you can use the package manager even in IDEs like Jupyter, as long as you first load the corresponding package, namely, *Pkg*.

One of the biggest advantages of Julia, however, is not its code-related aspects, but the people who make it shine. Namely, there is a team of talented developers who work to fix bugs and develop new features for this language. Note that this is not some random start-up but one that involves a core of MIT graduates, as well as a professor who works at the well-known technical institution. What's more, there are developers worldwide who contribute to Julia's code base, as well as enthusiasts who discover and report issues with the existing code. Finally, there are lots of people who use Julia in their work, be it financial modeling, scientific research, and data science projects. All this makes

Julia not just a language, but a vibrant community that continues to grow. The large number of Meetup groups for this language may also have contributed to this phenomenon.

Finally, Julia is also at the epicenter of deep learning-related innovation. From the deep learning package Knet to the Gen programming language (https://bit.ly/3b03M5W), Julia has been driving innovations in this AI-related area. Note that although Gen is branded as a wide scope AI language, it also handles probabilistic modeling, particularly Bayesian Statistics, while it also has an application focus in Computer Vision. MIT could have gone with any programming language, but they decided on Julia, something which is quite telling regarding the language's promising presence in our field. As for the Knet package for deep learning, it is the first deep learning framework written entirely in Julia, something that attests to the demand for alternatives to existing systems like MXNet and Tensorflow.

If all this hasn't convinced you of the numerous merits of the Julia language, perhaps you need to try it out for yourself. Many people who were hardcore R users shifted to Julia over the past few years, and there have been many other users who have formed a positive opinion of Julia after using it for a bit. Given how challenging it is to learn a new programming language (unless you are a coder already), this says a lot about the ease of use that Julia exhibits and its similarity to other high-level languages. Besides, there are numerous tutorials and introductory courses on the language, making it easy to learn. As for more advanced stuff regarding Julia, there is always this book!

The book is structured as follows. In chapter 1, we'll look at Julia as a programming language today and focus on what makes it stand out as a data science language, particularly on the Machine Learning front. In chapter 2, we'll explore how you can set up Julia on your computer or on the cloud. We'll also look into the IDE options in this chapter and explain why we delved into

Jupyter in this book. Chapter 3 will examine the essential libraries of Julia for data science work. Packages related to visuals, data structures, and some very useful mathematical processes, will be covered in this part of the book. In chapter 4 we'll shift gears and look at Machine Learning—there is a need to clear some things up before we look at specific ML libraries using Julia, which we'll do in chapter 5. Chapter 6 explores examples and exercises, and chapter 7 shows you how to code a Machine Learning model from scratch using Julia. To make it more interesting, the model we'll be working with is not one that can be found in any package, although it is one that has been proven to be effective and efficient.

Naturally, this book would be incomplete, if we didn't look into dimensionality reduction methods based on Machine Learning. That's something we'll delve into in chapter 8, through a closer examination of two of the most mature packages. In chapter 9, we'll examine some additional topics that play a role in Machine Learning, like parallelization and proper data engineering processes, which are oftentimes essential prerequisites to data modeling. In chapter 10, we'll look at how all this Machine Learning work impacts the business world and how to best use this know-how when liaising with other project stakeholders. We'll then examine some useful considerations to have about this subject in chapter 11, and we look at the future trends of Machine Learning and Julia in chapter 12. We conclude with some suggestions for what you can do next.

As a supplement to your learning of this subject, there will be a series of questions for most of the chapters, so that you can test your understanding of the topics presented. These questions can help you think about this subject critically and explore it in more depth, gaining a better understanding of Julia and its usefulness.

Beyond all this material, there is also an extensive glossary of the most important terms used in this book, as well as three appendices with supplementary material. The first appendix contains all the answers to the questions and exercises throughout the book. The second appendix looks at Julia's relationship with other programming languages and how you can leverage the bridge packages. Finally, the third appendix contains three heuristics, largely specialized but quite useful for data science work. All of them are implemented in Julia and are a good addition to the material presented previously, as they don't exist in any of the existing packages yet can add value to a data science project.

Naturally, since this is more of a hands-on book, it is accompanied by a series of Jupyter notebooks that contain all the examples presented in the text. On top of that, the data files of the datasets used, as well as any auxiliary scripts, are also provided. You can find all of this material at https://technicspub.com/julia. Note that future versions of Julia may not be 100% compatible with this code, so you may need to make alterations to it to ensure it works with the Julia version you are using. All the code here works well with Julia 1.1.

Upon reading and practicing the material presented in this book, you should be able to have a good grasp on Machine Learning and how you can leverage Julia for related projects. However, there is plenty more to learn on this topic, which is why this book is best seen as a starting point rather than a complete resource on the topic. After all, the field of Machine Learning is blooming, so there are new things coming up constantly. However, if you have a solid understanding of the topic, you should be able to learn the new methods and techniques quicker and better.

So, without any further ado, let's get started.

CHAPTER 1

Julia Today

Let's now look at how Julia fares as a programming language today. We'll start by briefly examining its programming paradigm, talk a bit about data science languages in general, and then look at Julia as a multi-purpose language. Next, we'll explore Julia as a data science language, and talk about the user community. We'll conclude by examining some useful resources for learning Julia basics.

The functional paradigm of programming

Programming can be done in various ways and today we have even more paradigms than the previous generation. Namely, we can choose among the following programming paradigms:

- procedural
- object-oriented
- functional

Procedural programming has to do with providing simple instructions to the computer (or any programmable machine) so that it can follow a particular procedure effectively. A procedural program contains a systematic order of commands, statements, and functions designed to complete a computational task or process. Languages like C, FORTRAN, Pascal, and Basic are procedural languages. Despite their long history in the computing world, some of these languages are still relevant today.

The key structures of procedural programming are the variable and the command. The former involves the storage of data and the latter its processing. Procedural programming is fast and fairly easy to learn, though not as easy to master since when doing complex tasks in a procedural language, it can take many lines of code. Procedural programming is often referred to as imperative programming. Despite its usefulness in computer science, procedural programming is rarely used in data science, unless it's a C script for a very specific supportive role.

As for object-oriented programming (OOP), this is the paradigm that involves the use of objects as the key structures of a program. An object is a software bundle that is characterized by states and behaviors. Much like physical objects, objects in OOP are defined by a series of properties and actions that can be taken related to these objects. This makes the OOP paradigm fairly intuitive and easy to work with, though mastering it is still somewhat time-consuming. After all, the plethora of data involved in these objects can be overwhelming and difficult to handle, especially if there are many objects that have similar attributes.

The actions related to the objects are represented by the various functions that can be applied to them. These are defined within the class that describes the most general form of an object. That's why we say that objects are instances of particular classes.

Let's look at a non-programming example, namely the car Fiat Punto with a particular license plate. This particular vehicle is an object of the class Fiat Punto, which describes all the cars of that Fiat make and of the Punto model. We can have a more general class called Fiat that describes all the vehicles and other products developed by this automotive company, thereby forming a hierarchy of classes. An object like a particular Fiat Punto can have functions like "drive", "park", and "fill up with gas." Each one of these functions applies to all the objects of this class, since other cars like this one can be driven, parked, and filled up with gas. However, an electric vehicle of the class Fiat Punto may have other functions, such as "charge battery" while the function "fill up with gas" wouldn't make any sense for cars belonging to that sub-class consisting of electric cars under the Fiat Punto umbrella.

Object-oriented programming is the most popular paradigm today and Julia follows that to some extent, since it allows for class structures and objects. Other languages that follow the OOP paradigm are Python, R, C++, C#, Java, and Javascript. However, debugging an OOP script can be time-consuming, particularly if it's complex, which is why a new paradigm was developed to alleviate this shortcoming of OOP: functional programming.

Functional programming is all about functions and their application on different inputs, to yield outputs. It's a fairly straightforward paradigm, though it has its limitations too. For example, a strictly functional programming script doesn't allow for a common workspace where variables can be shared among different functions. So creating more complex scripts can be a challenge, though this peculiarity of functional programming allows for faster execution speed and mitigates the chances of errors due to variable conflicts. Naturally, the scripts of a functional language are easier to debug and maintain, compared to those of other programming paradigms.

Julia follows the functional programming paradigm to a great extent, though it does allow for shared variables among different functions, bypassing this peculiarity of a nonexistent shared workspace. Also, the variables that are not used any more are discarded automatically (also known as garbage collection), which conserves memory resources and prevents memory leakage.

The functional programming paradigms are quite popular today due to performance. Although, many languages that make use of functional programming are also compatible with the OOP paradigm. A good example of such a language, other than Julia, is Scala. Additional languages that follow the functional paradigm include Haskell, Clojure, F#, and Elm.

Data science programming languages

What does all this have to do with data science and Machine Learning though? Few people are aware that many of the programming languages interact with each other. For performance reasons, it's often the case that a program is developed in one of them (e.g. Python) and implemented in another one (e.g. Java). This is known as the two-language problem, something that is a necessary evil when using a certain kind of language to prototype (namely a high-level language) and a different kind of language to deploy the program (namely a low-level language). The two-language problem creates additional latency and various kinds of liabilities such as coding errors, all of which can be avoided through the use of a language like Julia.

High-level languages are those that are easy to use since their syntax is closer to how we think and communicate. They are fairly easy to learn too, and because of their intuitive nature, developing a program in them can be fairly fast, even if you are not a professional programmer. High-level languages are key to data science since most data scientists don't care much about honing their programming skills, since there is so much more to learn in order to be able to work in this field. High-level languages include Python, R, MATLAB, and Julia. MATLAB is not used as much due to its high license fees, but you may see some people using its open-source clone, Octave. However, both MATLAB and Octave are fairly slow and despite their ease of use, they are not used as much in data science today (especially Octave, as it's not as practical for more complex problems due to its lack of packages). Other high-level languages are also somewhat slow, since the code written in them needs to be translated into something the computer can understand. Julia, however, is an exception to this, as we'll see later on.

Low-level languages are those programming languages that are "close to the metal" meaning that they are near the language computers understand: machine language. These languages don't need much translation for the computer to understand, which enables them to be super-fast to run. In fact, most performance critical processes, such as those responsible for infrastructure-related tasks, are handled by low-level languages for this reason. That's why programmers often spend years learning these languages, as these languages are not as intuitive as high-level languages. Some low-level languages (e.g. Java) are used in data science, but only as supplementary tools due to the challenge in using them to prototype. When using a high-level language however, it's not too difficult to migrate to a low-level language to ensure high performance.

Julia as a multi-purpose language

Let's now zoom in on Julia as a multi-purpose language—it follows the functional paradigm but is also connected to the OOP one. First of all, Julia was not designed to be a niche language, since much like most programming languages, it exists to bridge the user and the computer efficiently. Perhaps that's why there weren't that many data science packages in it when it was first released. It was only after some of us discovered that it can be leveraged as a programming language in data science.

Whatever the case, since it has reached a critical mass of users, Julia has been applied to many problems, ranging from the strictly scientific to the more applied. It became particularly popular in economics and finance at one point, so much so that the best introduction to Julia was a niche blog run by an economics professional. Also, a little known fact, many financial companies use Julia. After all, in domains like this one, performance matters. So, a high performance high-level language is like a dream come true.

Julia has developed in many ways, however, beyond economics and finance applications. For example, Julia has been successfully used in self-driving cars. Tiny computers (these computers that work on a single board and are very popular today among DIY tech enthusiasts) can run Julia too, since it's a cross-platform language, much like every other serious language. As a result, it doesn't take much imagination to figure out that it can be leveraged in computer vision applications which are an essential part of self-driving vehicles. Check out this video corresponding to this project, from a time before Julia v. 1.0: https://bit.ly/2JV8wh1.

The fact that Julia is so versatile makes it a useful platform for professionals and hobbyists alike. That's why there is a large variety of packages for all kinds of applications—see the official Julia website: https://pkg.julialang.org/docs. Currently there are over 2,000 packages for Julia and the ones on this webpage are just the official ones. There are a few unofficial packages too, plus nowadays it's easier than ever to create your own packages in Julia. So, even though Julia is often seen in the context of data analytics applications, it is much more than that.

Julia as a data science language

Let's now look at how Julia fares as a data science language and how it can be leveraged in the development and testing of Machine Learning models. First of all, although Julia was not designed with data science in mind, it ticks all the boxes of a data science language. Namely:

- it is easy to code and quite intuitive in its syntax
- its scripts are easy to debug and maintain
- it has great plotting packages enabling good data visualization
- it has several other packages that delve into Statistics and Machine Learning
- it liaises with other programming languages that are used in data science
- it is not esoteric and thereby it is accessible to people not trained in programming
- there is at least one good book on it, showcasing its usefulness in data science
- there are several companies that value Julia-savvy data scientists

As a bonus, it is fairly fast, so it overcomes the two-language problem. Not many data science programming languages can do that. As for the data science languages that are high performing (e.g. Scala), they are not nearly as easy to use as Julia.

But has it been used in practice for data science tasks? The short answer is yes, definitely. The longer answer is yes, but not as much as it should. This creates a bit of a discrepancy since many data scientists are either too busy or too conservative to give it a chance. After all, Python and Scala are good for data science work, so why would someone want to jump ship on them? Even people who have attended Julia Meetups and other events, showcasing how promising this language is in data science work, are still often undecided. The most common reason given is that they are not sure about the packages being enough and good enough. This is a big topic and it deserves its own chapter, which is why we are not going to go into it right now. Suffice to say that these concerns are not justified and that you can carry out data science projects using Julia exclusively.

Julia is great for data science because it is not tied to any particular platform, especially a cloud one. Even though there is an indisputable partnership with a major tech company, enabling Julia to be an option in that company's cloud, Julia can be used anywhere. Also, the fact that it is not widely known which company this is, shows that Julia is still independent.

Of course, Julia Computing has its own cloud service which it promotes for the seamless use of the language on a server, but they still give you a choice as to where you deploy your Julia programs. Other languages, like Swift for example, are tied to a particular ecosystem (in this case Apple's), which may be limiting in some cases. Julia doesn't do that and is extremely unlikely to do that in the future, either. All this enables Julia to tie in to all kinds of frameworks and tech ecosystems, making it a versatile piece of technology, compatible with many pipelines.

Julia as the epicenter of a community of users

Beyond the technical aspects of the language, it's good to be aware of the community aspect of it too. After all, it was never intended to be a niche tool for a few programmers having too much time on their hands! Instead, it was created to be adopted on a larger scale and aid in numerous areas, not just computer science. At least that's what can be derived by reading the thesis from MIT's Dr. Jeff Bezanson, who is one of the creators of the Julia language.

Although the Julia community is fairly small compared to other programming language communities, it is growing at a fast pace. Beyond the Julia conference that takes place annually (http://www.juliacon.org), there are several other events and groups where Julians gather. From the online ones such as the GitHub groups and Slack channels, to the face-to-face ones such as Meetup groups, the Julia community is vibrant. You can learn more about it at the official web page: https://julialang.org/community.

What's particularly interesting when it comes to Julia enthusiasts, is the diversity of the community. Julia may have started in a fairly technologically advanced corner of the world (Cambridge, Massachusetts), but it has spread to every part of the globe, including less developed areas. At the time of this writing, there are Meetups all over North America, Europe, Asia, Oceania, and even South America (particularly Brazil). What's more, there is a lot of work being done in Julia Computing to ensure that the set of users of the language remains diverse and as inclusive as possible.

Now, as mentioned previously, you don't need much help to pick up this language since it's fairly intuitive and there are plenty of educational resources at your disposal. However, for someone new to it, or someone who is not particularly versed in programming, it may be a challenging task. Having a support group of people who are in the same boat, as well as people more experienced in Julia programming can be a big plus. Participating in hackathons and talks may be enough to motivate or even inspire you to invest some time in learning and mastering this language.

The value of the community goes beyond all this, however. After all, just like most programming languages, Julia is an open source project. This means that even the source code of the language itself is open to everyone. Of course, not everyone is capable of adding something useful to the code-base of the base package (the core of the language), but everyone can report bugs, run tests, and even contribute additional packages. If you are more courageous, you can give direct feedback to its contributors and even promote Julia in an educational platform. Whatever the case, community members are encouraged to refine and promote the language in whatever way they feel comfortable. Perhaps this is why it has grown so quickly, unlike its closed-source counterparts, such as MATLAB, which although very useful, is limited to a niche user group, mainly researchers and engineers in large companies.

What may be most inspiring about the Julia community is that everyone there, including its creators, are very approachable. Some of these people are the equivalent of celebrities in the programming world, yet they are down-to-earth and happy to share their stories like you'd share stories with a friend over a coffee or a drink. Perhaps that's the most powerful aspect of the community and what makes it stand out from other programming language user groups, since it's akin to a group of friends who share a common interest.

Useful resources for learning Julia

Before we finish this chapter, let's look at some useful resources to brush up your Julia know-how. All of these are free, so you don't need to spend any money on learning the language. If you do have some funds that you'd like to devote to that purpose, feel free to buy my *Julia for Data Science* book!

First of all, there is a simple tutorial for v. 1.0 of the language (a fairly new release which is compatible with the one used in this book, v.1.1.1): https://bit.ly/3a3Bwhn. This 150 minute video, developed by the company that created Julia, covers the basics of the language, with plenty of examples to clarify all the points made. It won't make you an expert in Julia, but through it you can learn the ropes of the language and better understand the more advanced Julia resources.

In addition, there is the official documentation of Julia: https://bit.ly/2RsA1D6. This more esoteric resource may not be for the fainthearted, but it does contain all the information related to the functionality of the language that you'll ever need. Also, it's practically impossible to find a more reliable resource, since it was created by the makers of the language and the people committed to maintaining and evolving Julia.

Finally, Julia Wikibook is another great resource when learning the language: https://bit.ly/3b6JDLq. Contrary to other third party resources, this one is as robust as it gets, without being a Julia Computing resource. The Wikibook is not related to Wikipedia in any way, though it does use the same framework (wiki) as the well-known online encyclopedia. However, it reads like a good book and it is regularly updated. This is one of the few resources used in this book that is also used in the *Julia for Data Science* book from 2016.

Beyond these resources, there are plenty more, some more reliable than others. Note, however, that there is no golden resource that can make things easy for you all the way to the mastery of the language. If you are serious about learning the ins and outs of Julia—not just in theory but in practice, you need to use it in projects that are of real value to you, forcing yourself to become familiar with its more subtle aspects. Unfortunately, there is no book that can help in that, since the best way to truly learn a powerful tool like Julia, is to use it to solve challenging problems. Fortunately, before long, this whole process will become more rewarding and somewhat less challenging.

Useful considerations

There are several layers to a sophisticated programming platform and Julia is not an exception to this, no matter how exceptional it is as a language. For instance, Julia is bound to change, as every other programming language changes—except for the various legacy languages that have stopped being relevant in the 1990s. This means that many packages that work fine today may be unusable if they are not maintained to be compatible with the newer and better releases of the language. We'll talk about this more in the next chapter.

What's more, Julia is now a professional language, quite distinct from other new languages. A good example of such a language is Nim, an elegant and powerful programming language that although it has been around for over a decade, it's still under the radar for most data professionals. Nim may be a fairly good language, but it's still more of a novelty and not something someone would learn and hope to use for work projects related to data science.

Note that Julia scripts are not the most secure code files. After all, the programs written in Julia are not compiled, so in order to run them, you need to have the source code exposed. That's why it's best to shield them with APIs, when possible, if you want to have the option of third parties making use of them all while respecting the privacy of your code. Besides, due to the intuitive nature of the language, even someone who doesn't know it well can still figure out your code by looking at it—so you may want to make sure it's secure, especially if it involves critical processes in your organization.

Finally, Julia is not a solo player when it comes to the data science world, unlike Go, for example. Julia can collaborate well with other data science languages, such as Python and R, using the corresponding bridge packages. Also, there are packages that link Julia with C and Java, so you can always leverage packages of all these languages in your Julia scripts. What's more, if you are in a Python or R kernel, you can import Julia scripts too. All this makes the integration of Julia scripts in existing pipelines something doable—though if you are really concerned with performance, it would make sense to migrate your code base to Julia sooner or later.

Summary

- There are different programming paradigms, the most important of which are procedural (imperative), object-oriented (OOP), and functional.

- Julia is a hybrid programming language which although it follows the functional paradigm closely, it has a lot of elements from the OOP one, too.

- Data science programming languages tend to be high-level and, as a result, not particularly fast.

- Julia manages to combine the merits of a high-level language with those of a low-level one, solving the two-language problem—that is, needing both kinds of languages for a data science problem. Also, Julia is an excellent multi-purpose language, having applications that go beyond data science, as well as a respectable number of packages.

- The user community of Julia is a vibrant one and is characterized by diversity, breadth, and accessibility, among other things. Particularly, if you are new to the language, you can benefit from it in various ways, while if you believe in the language and wish to help refine or promote it, there are plenty of ways to do that too.

- There are various resources for learning the basics of Julia, some of which are free. However, the best way to master it is through practice, beyond programming tutorials and specialized books.

- There are certain things you need to consider about Julia as a programming language, such as the fact that it's constantly evolving, its professional presence in the world as a production-ready language, the need to keep your Julia scripts secure when using them with third parties if the source code is part of an organization's IP, and the fact that Julia can collaborate with other programming languages through bridge packages.

Questions

1. What makes Julia a functional language?

2. Why is Julia relevant as a data science language?

3. Do you need to be a programmer to learn and master the Julia language?

4. Isn't Julia too new to be reliable as a programming language?

5. How does Julia compare with Java in terms of performance?

CHAPTER 2

Setting up Julia

As with other programming languages, Julia is a program, so it needs to be installed and run like any other application on your computer or the cloud. That's something that hasn't received enough attention, and although the average developer won't have any issues handling this process, many data scientists may not find this process as intuitive.

Since we have more urgent things to do, going through the lengthy documentation of Julia to figure out how to make it run on our machines is not a priority. That's why it's good to have a straightforward way of installing Julia, namely a simple step-by-step guide. Also, as the use of IDEs greatly facilitates data science work, we need to install the IDEs and ensure they are working well with Julia. This task may prove even more challenging than installing Julia, since various key steps need to be taken in order to do this properly, something that is not clear in the instructions of these IDEs.

In this chapter, we'll start by looking at how you can install the kernel of the language—that is, install the actual programming language on your machine and then explore how you can do the same with the Jupyter IDE, which is the most popular IDE for data science work. We'll then look into other IDE options for Julia and where you can find them before we take a look at the JuliaBox possibility. Next we'll look at how you can handle Julia scripts and Jupyter notebooks for your work and examine some useful considerations to have about this topic.

Julia kernel installation

Let's start by looking at how you can install the Julia kernel on your machine. Fortunately, the Julia language is compatible with most operating systems, including Linux-based ones (e.g. Linux Mint) and FreeBSD. You can download the installation files here: https://julialang.org/downloads.

Unless you have a good reason to opt for the latest release of the language, you can make use of the latest stable release (usually the first one listed on the aforementioned web page). Then you need to find the most relevant file for your computer. If you are using a 64-bit machine that runs Windows, for example, you will need the .exe file that is on the right of the first row of the table. Note that most machines nowadays are 64-bit—unless you have an older computer.

You can find specific instructions on how you can install the Julia kernel for each operating system at https://julialang.org/downloads/platform.html. Note that if you want to uninstall the kernel for whatever reason, it's best to use the process provided at the end of the web page. If you decide to do that, however, make sure that you have kept your scripts elsewhere so that they are not deleted accidentally. You can replace the Julia kernel easily, but replacing your scripts may be next to impossible. Once you have installed the Julia kernel, you can execute the corresponding file and run Julia. For a Linux-based system, for example, you just type julia on the shell, as shown in Fig. 2.1.

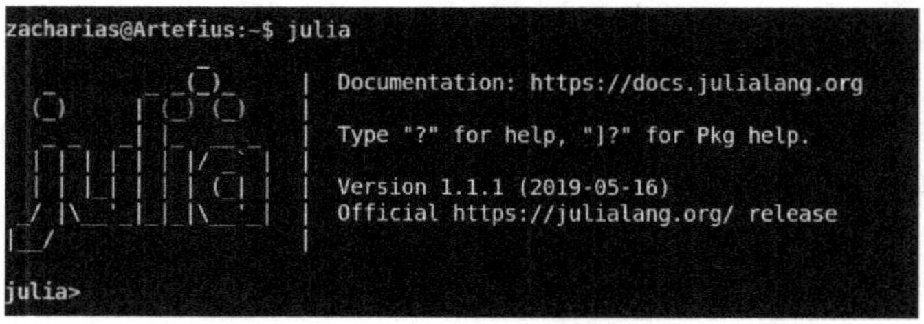

Figure 2.1. Screenshot of the Julia kernel being run on a Linux-based system.

Note that the prompt changes to julia> once the kernel is loaded on your computer. From this point onward, you can run Julia commands on your computer. To exit the language and return to the shell, you can either type exit() or press Control and D.

Jupyter IDE installation

Let's now explore the Jupyter IDE, which is also the best option for data science work, especially when it comes to wrapper methods and merging the various components of a program. Jupyter has been around for a while, but only fairly recently has it gained popularity thanks to its usefulness in data science projects. Whether you use Python, R, or Julia, Jupyter can provide a very useful and intuitive interface for your programming work, making any programming task more efficient and comprehensible, especially when dealing with complex tasks involving lots of commands and outputs.

The Jupyter IDE is a notebook where you can mix formatted text, code, and the output of that code neatly. The output can include graphics as well, making the notebook a very comprehensible demonstration of your work, which you can showcase at a data science meeting, for example. The fact that Jupyter allows you to export the whole notebook as a PDF file, for example, can help greatly in showcasing your work to a larger audience.

In essence, Jupyter comprises two main screens, the home screen and the notebook one, both of which are viewed on your browser through a local server the Jupyter program uses. The first screen, which you can view in Fig. 2.2, shows you the different notebooks on your machine and when they were last modified. You can also view any folders there are there and navigate through them, much like a file explorer. On this screen, you can also rename the notebooks and even delete them. Finally, you can see which notebooks are currently running. Clicking on any one of these notebooks will open it, usually in a new tab on your browser.

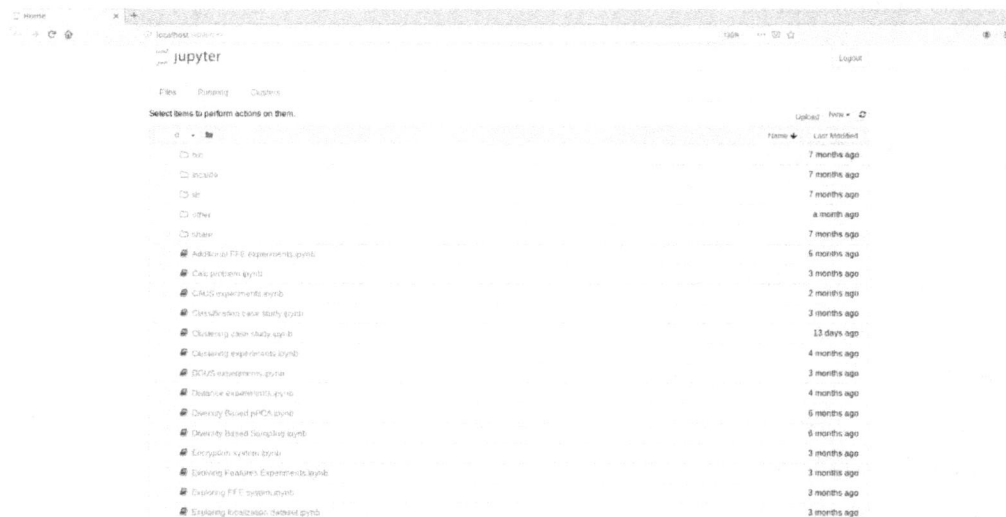

Figure 2.2. Screenshot of Jupyter's Home screen.

The Jupyter notebook is a bit more interesting. It looks like a document editor but allows for writing and executing code. All the lines that have been marked with In[#] are where executable code is included, while those marked with Out[#] are where the output of that code is shown (# is a number, related to the order in which that particular cell was executed). Note also that you can include headings to organize your work better and make the notebook easier to read. A sample of a Jupyter notebook appears in Fig. 2.3.

You can download and install Jupyter on your machine through Python by typing the following on the command prompt:

python3 -m pip install jupyter (for Python 3), or

python -m pip install jupyter (for Python 2)

Note that you may want to upgrade the pip program before installing Jupyter, by typing:

```
python3 -m pip install --upgrade pip
```

or

```
python -m pip install --upgrade pip
```

Figure 2.3. Screenshot of a Jupyter notebook.

To make Jupyter usable with Julia, however, you need to also install the IJulia package on your machine, through the Julia prompt:

```
Pkg.add("IJulia")
```

Unless IJulia is properly installed, you will not be able to use Julia on your Jupyter Notebook, even if the latter runs fine. So, make sure you pay attention to this step before proceeding to the next one.

To run the Jupyter program, you need to type on the command prompt:

jupyter notebook

Note that although Jupyter notebook is still the most popular IDE for data science work, there is also a new IDE developed by the same company, called JupyterLab. This is still fairly new, so it's a bit premature to know for sure its efficacy in data science work. However, the design seems to be slick and user-friendly, while the core functionality of the Jupyter Notebook remains more or less the same. In essence, the GUI is the only key difference, while the multitude of windows it entails makes it useful only if you are using at least two monitors. Not a bad option but perhaps a bit hyped overall since there are few things it offers that you cannot do on a conventional Jupyter notebook.

You can learn more about Jupyter Notebook and JupyterLab at the official website: https://jupyter.org.

Other IDE options

Julia has gained enough traction to be acknowledged as a relevant-enough language to have a presence in other IDEs as well. Also Juno is a special IDE designed for Julia specifically. Although it has been unreliable and does not work seamlessly in all operating systems, Juno is one of the first IDEs to accompany the programming language. After all, when it does work, it is very practical and ideal for low-level programming in Julia. Also, before it became its own thing, it was part of Atom, another useful IDE for the language. Note that you can use Atom as an IDE, regardless of Juno. The former is quite stable and works well with Julia.

Atom is probably the best IDE for programming in general, covering a large variety of programming languages, while it can also be used to view text files efficiently. Lately, it includes Julia scripts too, though getting the Julia kernel to run on Atom can be quite a challenge, especially on a Linux-based operating system. However, you can still get the IDE to recognize Julia code and show it with some colors, something that can make every Julia script more comprehensible. You just need to make sure that the package language-julia is installed on Atom for this to happen. Note that Atom has a bunch of other useful packages you may want to install as well, though there are no other Julia-related packages that you need for using Atom as an IDE.

Other IDEs include Vim and SublimeText, both of which are exceptional text editors and not just for editing scripts. Note that beyond these editors, any other text editor can function as an IDE. However, without the Julia IDEs syntax formatting dictionary, it would be difficult to read the code properly, while debugging may be more time-consuming. Finally, there is a reference somewhere to Weave as an IDE. However, this is a misconception since Weave is a scientific report generator package designed by the same group that developed Juno, namely JunoLab.

JuliaBox

JuliaBox is another interesting option for running Julia, especially if you need to access the language remotely. Note that this is primarily a commercial option managed by the Julia Computing company, so if you are serious about using it, it would be best to think about your budget. Naturally, there is a free option so that you can try it out. In Fig. 2.4 you can see a screenshot of what it looks like right after the login, while in Fig. 2.5, you can view it after the Jupyter app is launched. As you can see, it looks very much like running Jupyter Notebook on your machine. Note, however, that the versions of Julia that it supports are not the latest ones. Yet, it does support multi-threading, so if you want to use the full spectrum of computing resources in it, you have that option, even for the free option.

Figure 2.4. Screenshot of the JuliaBox system, after login.

Figure 2.5. Screenshot of the JuliaBox system, after launching Jupyter. Note that the files system of every JuliaBox account comes equipped with some tutorials for the newcomers of the language.

Although JuliaBox is fairly popular among many Julia learners, particularly in universities, its main use case is deploying Julia programs in need of a large number of resources. So, in order to make the most of it, it is best to invest some money in it, though preferably after you have implemented and tested your scripts. This way, you can use it without having to worry too much about potential bugs in your code, something that can cause frustrating delays, which you would have to pay for if you have to do the debugging of your scripts on the Julia cloud.

Nevertheless, JuliaBox is the last piece of the puzzle in the data science pipeline, since it can enable you to do even the most challenging part of it, model deployment, in Julia, without having to rely on a tool from a different ecosystem. This wasn't the case a few years ago when Julia first began to make a case for data science and Machine Learning applications. And even if it's not too difficult to create an API in Julia, JuliaBox makes the whole process much easier for deploying your programs on an internet server.

Note that JuliaBox is not the only cloud option for Julia. You can run Julia on the Microsoft cloud (Azure), something you can learn more about in this article: https://bit.ly/3eghymM. Of course, such a move would make sense if you already have worked with Azure or wish to deploy a Julia script in combination with a cloud-based database or other programs that dwell there. For most data science applications, the JuliaBox option would be more than adequate.

Julia scripts and Jupyter notebooks

Handling Julia scripts and Jupyter notebooks is something few people will talk about as it seems obvious to them, even if it doesn't seem so straightforward to someone new to the language or to how Julia is used in a data science context.

First of all, Julia scripts are the .jl files that contain Julia code. Any such file will be recognized as having Julia language syntax by an IDE that supports Julia. This way, you can open them and view them in an understandable manner. Note that even if the extension is different, the Julia scripts you have in text files would still run on Julia. However, it's generally good practice to store all your Julia programs in .jl files. A key .jl file you need to be aware of is the startup.jl script, which is located in the /etc/julia subfolder of the folder Julia has been installed. So, if you want to execute certain commands as soon as Julia starts, that's where you'd put them. For example, you may want certain libraries to be loaded or change the working directory to a particular folder. You can do these things easily by adding the corresponding commands in the startup.jl script. Note that this may slow down Julia a bit, so it's good to remember where this file is kept and perhaps make a backup of it before you modify this script.

As for the Jupyter notebooks, these are the .ipynb files. This bizarre extension stems from IPython Notebook, as they were used primarily for Python scripts. As the makers of the Jupyter notebook platform would tell you, Jupyter was created primarily for three programming languages: Julia, Python, and R, which is where its name comes from (Ju from Julia, Py from Python, and r for R, with the te part as a filler syllable).

Anyway, Jupyter notebooks are also text files that are designed to be used by the Jupyter program (technically they are JSON files, so if you were to open such a file in a text editor like Atom, be sure to use the corresponding package to parse them properly). Any project you make using Julia is best stored in such files, though you may use .jl files in it too, preferably stored separately, so that they can be leveraged in other projects also. Note that for .jl files that you plan to use in a different project, it's advisable to structure all of the code in them as functions.

For larger projects, you'd have several files, be it .jl or .ipynb ones. It is a good practice to store all of these files in the same folder, particularly a folder dedicated to that project. For .jl files that are used in several projects, it's best to have them in a general folder for easier access. Naturally, you'd want to give all the files of your projects intuitive names or names that you can easily remember. Otherwise, they are bound to get lost as you create additional files in your codebase. Also, for the most important Julia files you have, it's best to back them up regularly.

What's more, if you make changes to these scripts regularly, or if you work on these scripts in tandem with other people, you may want to use a version control system like Git or SVN. Also, as you may have noticed, the bulk of the code base of Julia is on Github, which is also the version control system most Julians use. You can use whatever version control system you like for your Julia work since they are about the same in terms of functionality and ease of use.

If you have lots of Julia scripts working together for a particular task, especially for a generic task, it's best to compile them into a package. Use the Pkg.jl package, a package for packages (a meta-package if you will). We'll look into this more in the next chapter, where we'll talk about Julia packages in general, focusing on the most important ones for data science projects.

It's a good practice to separate the scripts that you plan to use in various projects from the ones that are created for drills or one-time usage. Also, it's good to maintain the scripts you plan to use in the future, as the Julia language evolves from release to release, to ensure that they are always relevant and efficient. Bundling the most similar into larger scripts also makes sense as it makes it easier to manage them.

Useful considerations

Installing Julia and an IDE may be fairly straightforward, but there are certain things that you need to keep in mind nevertheless. For example, it's not uncommon to have multiple IDEs when using the language (or any programming language for that matter). That's something that we'd recommend, since the Atom IDE, for example, is much better at handling multiple scripts than a Jupyter notebook. However, you'll still need the latter if you want to do some data science work in a manner that enables easy collaboration and presentation. However, if you develop a new method that you plan to use in various projects, it's best to do that in another IDE, not Jupyter.

Also, when upgrading the Julia kernel, you need to be careful to ensure that the new release is working properly. If you just download the new release on your machine and run it, things may be fine if you do so from its folder, but the symbolic link to the new executable needs to be updated too. This omission may create serious confusion and make you waste a lot of time. Besides, once you get a grip on the latest release of the language, there is little point in keeping the older release around, unless you have legacy code lingering on your computer.

What's more, you need to be mindful of the package situation when running a different release of the language. After all, every new release is clean, and you have to install all the packages you need from scratch. It's not too challenging a process, but it does take some time, so you need to plan properly. Fortunately, since version 1.0 of Julia, the package installing and updating process is a breeze, significantly better than that of other data science programming languages.

Furthermore, it's important to check if a package you plan to use is working properly in the new release before you upgrade your Julia kernel. Otherwise, you may find that you need to revert to a previous release just so that you can run that package properly. Luckily, this doesn't happen often and when it does happen, it is usually a temporary issue. However, if you have crucial code that relies on a certain package, that's something you need to be aware of as it can save you time and effort.

Finally, all these IDEs and accessories of the Julia ecosystem may be great, but they are no substitute for effective programming and good practices, qualities that are both transferable and valuable when it comes to carrying out any data science project.

Summary

- The installation of the Julia kernel is a fairly straightforward process as long as you obtain the right file for your computer and follow the corresponding instructions.

- Installing an IDE for Julia is not essential but particularly useful, especially for data science work.

- Jupyter is the most popular IDE for data science projects and works seamlessly with Julia. Always install the IJulia package on Julia before trying to run Jupyter in combination with Julia, though.

- JupyterLab is another product by the same organization that developed Jupyter Notebook, sharing most of its traits but with a slicker interface. However, whether it improves the efficacy of data science work is something yet to be determined.

- Several other IDEs are compatible with Julia, such as Atom and Juno.

- JuliaBox is a popular cloud-based option for Julia, essentially a Jupyter notebook that you can access from any computer. Beyond JuliaBox, there is the Azure cloud option, too, for more niche use cases.

- Handling Julia scripts and notebooks is needed to maintain a certain security level for your work. It's best to keep all of these files in a separate folder, different than the one where Julia is installed. Also, regular backups of these files are highly recommended.

Questions

1. When would you use an IDE like Atom in your work?

2. Can you use a basic text editor for creating or editing your Julia scripts?

3. What kind of file would you use to store the code of a multi-purpose function you have created, to use in your data science projects?

4. Can you use Jupyter offline?

5. Is Jupyter better than other IDEs?

6. Can you run Julia on your mobile device?

CHAPTER 3

Julia Libraries

Let's now examine some key libraries for Julia that are useful in all kinds of data analytics projects. Libraries are usually referred to as packages, and in Julia, they are handled by a package manager program, which is part of the main language. Since there are plenty of packages now, it's good to prioritize which ones to use since some are better than others for a particular task. Also, as the language matures, some packages are left behind, and even if they were great once, they might not be so relevant anymore.

In this chapter, we'll look into packages for Julia, with a focus on functionality. We'll look at what packages are available in the Julia ecosystem and organize them into meaningful categories for a data scientist. Then, we'll look specifically at the packages related to data preparation (making the data ready for your models), data models, Statistics, Machine Learning utilities, AI, and some other packages that don't fit into any one of these categories (e.g. plotting packages).

A library of libraries

Fortunately, there are plenty of packages available in the Julia ecosystem for data science tasks. Contrary to what the critics of the languages say, nowadays, there is a package for everything you'd need in your data science work, even the most advanced models, as can be seen in Fig. 3.1. If you are comfortable with using mind-maps, you can use this diagram as a starting point to help you organize the packages you learn in a way that makes sense and is not too difficult to remember. Note that not all of these packages would be necessary for your data science work. Sometimes even a handful of packages suffice for a given project, while other times, it's easier to code something from scratch and use your scripts for your work. Whatever the case, knowing that these packages exist can be a useful aid, particularly when you are new to programming or the Julia language.

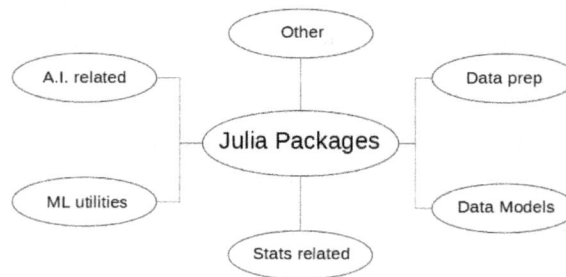

Figure 3.1. Main types of libraries in Julia languages, related to data science work.

For additional packages, the best place to start would be the official package repository of Julia: https://pkg.julialang.org/docs. Although not all packages are covered, the repository includes the ones that have been tried and tested enough for the Julia Computing company to give its stamp of approval. Also, these are the packages that the Julia community uses and provides feedback, so they are the most likely ones to be in good shape.

Data preparation libraries

Let's start our exploration of the Julia packages with the ones geared toward data preparation. Data preparation is the part of the data science pipeline that involves getting the data in a neat and usable state so that it can be fed into models, be it for exploring or for making predictions. The key data preparation packages in Julia are:

- **DataFrames** – the Julia equivalent of Python's Pandas package, and one that plays an instrumental role in data science tasks as it contains the Data Frame structure. Although not always essential for a data science project, it is very useful, and you can't do much without it if you plan to work with most of the data science packages since they have this package as a dependency.

- **ExcelReaders** – a package for accessing .xlsx documents and storing their contents into a data frame. This package does not create .xlsx files, however.

- **CategoricalArrays** – a package for handling categorical data, be it nominal or ordinal. The data may contain missing values too.

- **CSV** – a package for handling CSV files easily. Although you could create all the functions of this package yourself without being adept at programming, it's still a convenience when dealing with tabular data files.

Note that there are more packages that qualify as members of the data preparation family, but these are sufficient for most tasks. Besides, for more niche applications, you can find whatever package you need in the Julia package ecosystem. Also, all of the previous packages are available on GitHub, too, so you can access them using a GitHub account. However, it's easier to use them through Julia, either through the REPL or one of its IDEs.

Libraries for data models

Let's now look at some libraries geared toward data models. We won't go into much depth in them as they are going to be examined in some detail at a later chapter, along with examples. Depending on the methodology, they are most relevant to these packages:

- **Unsupervised Learning** – not making use of a target variable, such as for exploratory data analysis (EDA).

- **Supervised Learning** – making use of a target variable, be it for classification or regression problems.

- **Transparent models** – models that are easy to interpret and understand their functionality.

- **Black box models** – models that are difficult or impossible to interpret (figure out how they arrive at their conclusions exactly).

Based on this simple taxonomy, we can organize the data model packages as:

- Unsupervised Learning
  - **Clustering** – a package specializing in clustering methods for grouping data into meaningful groups, usually part of Exploratory Data Analysis.
  - **ManifoldLearning** – a package focusing on non-linear methods for reducing the dimensionality of a dataset.

- Supervised Learning (transparent model)
  - **DecisionTree** – decision trees and random forests package.
  - **GLM** – generalized linear model package for regression analysis.

- Supervised Learning (black box models)
  - **XGBoost** – a gradient boosting framework for ensembles using boosting, particularly comprising of decision trees.
  - **LIBSVM** – a package for Support Vector Machines (SVMs) in Julia.

Beyond these packages, there is another one that, although clearly under the supervised learning umbrella, it is difficult to classify using this taxonomy. The reason is that it contains models that span across both categories: the MLJ package and its sibling package, MLJModels. For reasons that go beyond the scope of this book, these two packages are often used in tandem.

MLJ and MLJModels are two packages used for various supervised learning models, spanning from simpler ones such as kNN, to more complex models such as Ridge Regressors, Learning Networks, and Ensembles. Although MLJModels is an extension of MLJ, MLJ is a unique data modeling package in Julia as it attempts to unify all existing models and methods within a single framework. However, as it's relatively new, it has some wrinkles that need to be ironed out before it can be considered production-ready. You can check it out at https://bit.ly/34swGZZ.

Note that it's best to have a decent understanding of the models behind the methods in these packages since the examples in the corresponding GitHub pages are lean and simple, perhaps too simple for someone unfamiliar with all this. However, this is not a criticism for these web pages or the people maintaining them, but rather an observation regarding the oversimplification of the data science field and the dangers this entails. We'll look into this in more detail in the next chapter.

Statistics libraries

Statistics packages are some of the most mature packages in Julia and the most widely used in various data analytics projects. The most important and reliable such packages are:

- **StatsBase** – a collection of multiple Statistics functions, essential in most data science projects. Probably one of the oldest packages in the Julia repository.

- **Distributions** – a package for all kinds of probability distributions, along with useful functions for performing various tasks with them, such as sampling, maximum likelihood estimation, and the *pdf* of each (as well as its logpdf). See the package's official documentation for more information: https://bit.ly/3b4vz58.

- **HypothesisTests** – a package containing several statistical methods for hypothesis testing, such as t-tests and chi-square tests.

- **MultivariateStats** – a great place to obtain various useful Statistics functions, including *principal components analysis* (PCA).

Probably the best place to start would be with the distributions package, as it's often used as a dependency in other data science packages. Whatever the case, each one of the Statistics packages here is worth learning, particularly if you have a sufficient understanding of Stats. Note that the GLM package would also fit in this category, but it seems more relevant as a data model package. After all, this taxonomy is used to facilitate the organization of the packages into a meaningful and memorable manner—to optimize their usefulness in data science projects.

Libraries related to ML utilities

Beyond Stats and model-related packages, there are some packages geared toward helping out with Machine Learning tasks. These are:

- **Distances** – a quite useful package for distance calculation, covering all major distance metrics—particularly essential for transductive systems, be it clustering algorithms or distance-based classifiers and regressors.

- **MLJ** – beyond Machine Learning models, this package includes a variety of utilities, too, such as grid search and K-fold cross-validation. Even if you don't use the MLJModels package, MLJ is useful in and of itself for data science projects.

- **ScikitLearn** – the well-known master package for all kinds of Machine Learning applications in Python, made available through the PyCall bridge package.

- **MLLabelUtils** – a quite useful package for pre- and post-processing of labels data for classification problems. Labels encoding may seem like a trivial task, but it can be time-consuming and prone to errors, something this package aims to help manage efficiently.

- **MLBase** – this is the most established package in Julia, related to Machine Learning, covering a wide range of tasks. You can obtain more information about it at https://bit.ly/2wyJmln.

- **AutoMLPipeline (AMLP)** – a creative approach to building machine learning pipeline structures, for more complex tasks. This tool can bring a great deal of value to a data science project. More information about it at https://bit.ly/2zLM6gt.

Although MLJ is the most promising, it's still not there yet, so we will focus on the other ML packages in this book. This package may someday be the Julia equivalent of Scikit-learn (even if the latter exists in Julia too as an API for the Python package). Still, until then, it's recommended to rely on the existing packages for any data science work.

AI-related libraries

AI-related packages aim to bring forward the more advanced models that are employed in data science, via Machine Learning methodologies. There are several, but the most important of them are:

- **Knet** – this deep learning framework is the first one that's been developed entirely in Julia (even the cost function methods are coded in the language). Created by Professor Deniz Yuret and his team, it covers a variety of ANNs, including MLPs, CNNs, and some models specializing in sentiment analysis.

- **Flux** – although this package is marketed for Machine Learning, it is one of the most well-made deep learning packages, developed using Julia exclusively. It also has GPU and AD support and is optimized for performance. Also, it is fully compatible with the Metalhead package. You can learn more about Flux at https://bit.ly/2RysQZX.

- **Metalhead** – one of the most creatively named packages in the Julia ecosystem, Metalhead is all about Computer Vision. So, if you are looking to perform classification on images using AI, this is the best place to start. More info at https://bit.ly/2Vg3IYQ.

Other AI-related packages in Julia corresponding to well-known frameworks such as MXNet are used with languages like Python and R. However, as this is not a book on these frameworks, we'll focus on the conventional Machine Learning packages instead. Besides, these are more promising and worth looking into, no matter how popular the more advanced AI-related libraries are these days.

Other libraries

Beyond the previous packages, there are a few more that are equally useful, if not more useful. Namely, this generic group of packages includes those related to specialized processes like graph analytics and of course plotting (data visualization). Also, there is a package that ensures that all packages in Julia work in harmony and are up-to-date. These packages are:

- **TSne** – a package implementing the T-SNE algorithm by the creators of the algorithm. Its creator and his team cover it at https://bit.ly/2ycKX0w. Note the stochastic nature of this algorithm and its scope, since this is not meant for everyday dimensionality reduction, like PCA and ICA, for example.

- **UMAP** – a package related to the Uniform Manifold Approximation and Projection algorithm, a powerful dimensionality reduction method. You can learn more about it at https://bit.ly/3ccLLkV. Just like T-SNE, UMAP is stochastic, but unlike T-SNE, it's much more useful in dimensionality reduction that goes beyond visualization. UMAP is often used in conjunction with Clustering methods.

- **Graphs** – the most complete graph analytics package. Note that these are mathematical graphs, and although there is a visualization aspect to this package, it is not relevant to plotting data in general, just graph-related data.

- **Gadfly** – one of the best plotting packages, written entirely in Julia. Note that the inputs of the corresponding methods require the use of the DataFrames package, mentioned in a previous section of this chapter.

- **PyPlot** – a great plotting package, borrowed from Python's Matplotlib, ideal for heat maps, among other plot types. If you are new to Julia and already familiar with Python, this is a good place to start in your visualization endeavors.

- **Pkg** – a powerful package for creating, installing, updating, and uninstalling packages in your Julia kernel. Learn more about it here: https://bit.ly/3efliF4. The most common usage of it is:

Pkg.add(CoolPackage)

Pkg.update()

Pkg.rm(CoolPackage)

Pkg.status()

Although the above format is valid, it is rarely used any more since this package is so common that a particular shortcut is employed instead. Namely, you can press the "]" key while you are on the REPL and gain access to it directly, so that you can run the above commands in a simpler format, as shown in Fig. 3.2. You can exit this mode by pressing the "backspace" key while you are on the Pkg prompt. Note that you need to import this package in a Jupyter notebook if you plan to add, remove, or update a package while in that environment. After all, the "]" shortcut only works on the REPL.

The upgraded usage of this package is one of the most important improvements of Julia since version 1.0. You should familiarize yourself with this package before learning any other package in the language, especially if you plan to use packages extensively for your work. Browse the many more packages available on the Julia package website.

```
(v1.1) pkg> status
    Status `~/.julia/environments/v1.1/Project.toml`
  [7073ff75] IJulia v1.19.0

(v1.1) pkg> update
  Updating registry at `~/.julia/registries/General`
  Updating git-repo `https://github.com/JuliaRegistries/General.git`
Resolving package versions...
  Updating `~/.julia/environments/v1.1/Project.toml`
 [no changes]
  Updating `~/.julia/environments/v1.1/Manifest.toml`
 [no changes]

(v1.1) pkg>
```

Figure 3.2. The Pkg package in action. You can enter this mode of usage by pressing the "]" key, while to exit it, you just need to press "backspace" while on the Pkg prompt. When actually updating packages, there is more text on the screen, depicting the progress of the update for each of the packages being updated.

Useful considerations

Although the package ecosystem of the Julia language is quite straightforward, some things are important to keep in mind to make the most of it and avoid any potential issues. First of all, Julia packages are always work in progress, and they rely on the feedback of users like you to remain relevant and useful as tools in data science work.

What's more, not all packages you come across in Julia are going to be fully compatible with each other. Take LightGraphs, for example, a package that aims to facilitate Graph Analytics tasks swiftly and intuitively, explained in my previous book on Julia. As great as this package may be, it is not compatible at all with the Graphs package, which is another, more well-established library of Graph Analytics methods.

In addition, packages are the product of some people who took time out of their days to create them, so they have their limitations, and may not be bug-free. They are not always the product of a team with sufficient resources to develop them in the professional manner you may be used to if you are coming to Julia from other more established programming languages. Take ELM.jl, for example, the package that put Extreme Learning Machines on the map for those data scientists who are unfamiliar with this kind of network-based system. The ELM package doesn't cover the various types of ELMs, which are what make ELMs such an intriguing technology.

Furthermore, Julia packages are a great facilitator of data analytics tasks, but they are no substitute for proper planning and meticulous application of the data science mindset. They may enable you to do interesting things and come off as someone who knows data science, but unless you have an understanding of the field that goes beyond methods and techniques, you are bound to remain at the surface and not fulfill your potential.

Finally, even if you are not an adept programmer or a Julia connoisseur, you can still take a stab at developing your tools using this language. They may not be optimal, and they may not be promoted as much by the more experienced Julians. Still, if they help you improve your efficiency at your data science work, they may be valuable regardless. If you are brave enough and have no copyright restrictions, you can even share your work with the Julia community and invite others to help you with these scripts, gradually evolving them into new packages which benefit everyone.

Summary

- Packages in Julia are sufficient in number and diversity, for facilitating various data science-related tasks, particularly those geared toward Machine Learning.

- There are different groupings of data science-related packages in Julia, such as the one proposed in this chapter: Data preparation, Data models, Statistics, ML utilities, AI-related, and Other. Whether this taxonomy is comprehensive enough or not, it can still be useful for organizing the various data science related packages in the Julia ecosystem and finding them faster when in need of them.

- Data preparation packages include DataFrames, ExcelReaders, CategoricalArrays, and CSV.

- Data model packages include Clustering, DecisionTree, GLM, XGBoost, LIBSVM, and MLJ/MLJModels. These cover a variety of models, ranging from the unsupervised learning ones to the supervised learning ones. Some of these correspond to transparent models (GLM and DecisionTree packages) while others to black box ones (ELM, XGBoost, and LIBSVM). The MLJ and MLJModels packages, which are usually used together, cover a broad spectrum of packages across different categories.

- Statistics packages include StatsBase, Distributions, HypothesisTests, and MultivariateStats. Of these four packages, the one that is most relevant and worth learning first is the Distributions one, though all of them are important. Some understanding of Statistics would be crucial for making the most of these packages, however, and for understanding the results of the methods they contain.

- ML Utilities packages include Distances, MLJ, ScikitLearn, MLLabelUtils, and MLBase. Of these, MLJ is probably the newest one and attempts to provide a unified framework for all the Machine Learning models available in Julia. ScikitLearn is more like a proxy for the well-known scikit-learn package in Python, while the MLLabelUtils package is useful for processing the labels data for classification-related projects.

- AI-related packages include Knet, Flux, and Metalhead. Of these, Knet is a deep learning framework comparable to the conventional ones used with Python, while Metalhead is specialized in computer vision applications.

- Other packages include T-Sne, Graphs, Gadfly, PyPlot, and Pkg. The most important of these are Gadfly (best plotting package in Julia) and Pkg (responsible for adding, updating, removing, and creating packages in the Julia installation you are using).

Questions

1. What's the point of using pre-made packages in Julia if we can code everything from scratch without any serious compromise in performance?

2. Can you trust 3rd party packages for data science applications in Julia?

3. What's the best data science-related package in Julia?

4. What would you do if none of the packages presented in this chapter are suitable for your data science-related project?

5. Do you need to know the algorithms behind the programs in the data science related packages to use them?

6. What can you do if you come across an issue (bug) in one of the official packages of the Julia ecosystem?

7. Do you need all of the packages presented in this chapter for your data science work?

CHAPTER 4

What Machine Learning is and isn't

With Machine Learning being such a buzzword nowadays, it's good to know about it and know it well. In this chapter, we'll cover the two main paradigms of data analytics and examine where Machine Learning fits. Following this, we'll proceed to debunk some myths about the subject, examine the main types of Machine Learning briefly, and then explore how it relates to AI.

Two main data analytics paradigms

Data analytics, the super-set of data science, has been around for a long time, even before data mining, the precursor of data science, was a thing. Back then, it was all about Statistics, but even then, there were elements of a different approach to analyzing data, an approach few people trained in Statistics know well enough.

Conventional Statistics makes use of mathematical models, particularly models related to distributions, to understand the data and make predictions. This is why it is often referred to as the model-driven approach. It's not ideal, since there are many assumptions Stats professionals make to use this approach, but it works adequately well, for many datasets. However, it is possible to analyze data without using such models, something that is explored to some extent with non-parametric Statistics. This approach relies primarily on the data itself and is therefore referred to as data-driven.

Many methods use the data-driven approach to data analytics. The catch is that to do something meaningful with this approach, you need to have sufficient data and clever algorithms to process it, something quite possible these days. After all, despite the impressive sophistication of Statistics, the models it offers are fairly simple. This simplicity of Statistics is what has contributed to its popularity. Even non-experts in the field can be trained in this subject and master it to the extent that they can make something useful. Also, the simplicity of Statistics when it comes to modeling enables transparency and ease of interpretation of the results these methods yield.

However, this is not a book about Statistics, so we'll focus more on the data-driven approach, which relies primarily on Machine Learning methods. Nevertheless, thinking that Machine Learning is the only way to go and that Statistics is not useful in data analytics is a dangerous idea, which is why there is a lot of emphasis in many books (including this one) on the usefulness of both data analytics paradigms. Even if the data-driven approach prevails today, it's good to acknowledge the value in the model-driven approach, since hybrid approaches also exist, combining the merits of both of these approaches. One such hybrid approach is Bayesian Statistics.

Machine Learning as a data-driven approach

Machine Learning is a set of algorithms and programs that aim to process data without relying on statistical methods. Machine Learning is generally faster, and some methods of it are significantly more accurate than the corresponding statistical ones. At the same time, the assumptions they make about the data are fewer, and in some cases, non-existent. This reliance on the data itself rather than some arbitrary mathematical model describing the variables involved is what renders Machine Learning a data-driven approach.

A few years ago, Machine Learning researchers had to prove that their methods were good enough and fast enough compared to the statistical ones. Often they relied on Statistics theory to drive home the point that they were truly scientific since there was little theory around Machine Learning. Naturally, the Statistics establishment fought Machine Learning harshly, until eventually the Machine Learning methods were proven to be at least as good, beyond any doubt. People started branding themselves as Machine Learning researchers and eventually earned the respect of professionals of other fields in the data analytics community. One of these fields was AI, an already established sub-field of computer science that was focused more on real-world applications, such as robotics.

It's important to note that the Machine Learning community and the AI one were distinct and remain separate to some extent. There are AI professionals who have nothing to do with Machine Learning and Machine Learning people who don't know the first thing about AI. However, there is an overlap between the two, something that began when Machine Learning adopted certain AI-based methods for predictive analytics. We'll look into that in more detail in a later section of this chapter.

Predictive analytics is data analytics that has to do with predictions (no surprise). In the Statistics realm, this is inference modeling, and it's probably the most important part of that field. In Machine Learning, most methods are created for predictive analytics. However, there are a few that are related to auxiliary processes such as feature selection, feature fusion, and feature evaluation. However, whatever Machine Learning is involved in is data-driven.

Some models make assumptions regarding the independence of the features involved, but most of them don't care. Nevertheless, the preparation of the data through cleaning, normalization, and other processes of data engineering, is still an important part, even if you rely on Machine Learning methods. Few methods can work with the raw data directly, while most of them rely on refined data for best results.

Machine Learning and the data-driven approach in general flourish because of the abundance of data available, the technology used is advanced enough and affordable, and the fact that people are more open-minded regarding the tools used for analyzing data. Nevertheless, most Machine Learning models tend to be black boxes. This means that, in general, you don't find in them the transparency and interpretability found in statistical models. However, there are exceptions, such as Decision Trees and their derivative models.

Machine Learning myths

Machine Learning has gained a lot of popularity lately, which has led to the creation of many myths and misconceptions.

For example, many people think that Machine Learning is directly linked to Statistics, some going so far as to see it as an extension of Statistics. This is as ridiculous as saying that medicine is an extension of physics, or that architecture is just painting but with ruler and compass. Take k Nearest Neighbor (kNN), for example, the simplest Machine Learning algorithm, dating back to the 1950s. Even though it has been analyzed in-depth since it came about, with some methods of analysis related to Statistics, it is in its core a data-driven approach that's under the Machine Learning umbrella.

Naturally, some people like to view the confidence scores that many algorithms yield as probabilities. That's fine since, in many cases, such a perspective helps us interpret this output of the models. However, just because we choose to view it as probabilities doesn't make the whole model probabilistic. That's a subtle but important point. Someone can, for example, say that she is 90% confident that the answer to the question is A and only 10% confident that it is B, as a way to quantify how sure she is about this choice. This may help us interpret what she believes and compare it with what other people believe for particular questions. However, how she arrived at this conclusion may have nothing to do with probabilities. You can view KNN's case similarly.

What's more, Machine Learning is not the same as AI. For starters, AI has been an independent field for longer, and it is more concerned about systems that exhibit intelligence, in any one of its forms, mostly computer-related ones. Machine Learning, on the other hand, cares more about understanding and processing data and making predictions. We'll explore this in more detail in the next section of this chapter and in the chapters that follow.

In addition, Machine Learning is not something you can pick up in a boot camp or by learning the documentation of a Machine Learning package. It's much more than that since it's a whole framework, spanning over a large variety of methods. Also, it's the whole mindset behind this and the problem-solving ability that is linked to all this. A good Machine Learning professional will do more than just build a relevant model from this framework. He may also need to sort out the data, talk to other stakeholders of the project, try out different approaches, decide on which one is best for that particular project, and present it in a coherent and comprehensible manner. What's more, a Machine Learning expert is bound to be able to learn new things on her own, while in some cases, she may even come up with some original Machine Learning methods.

Finally, Machine Learning is more than just the two to three methodologies most commonly used. Most books, including this one, focus on a subset of the various Machine Learning methodologies, for practical purposes, but there is more to it than that. For example, there is also Reinforcement Learning, Semi-supervised Learning, Self-supervised Learning, just to name a few. What's more,

Machine Learning is a constantly evolving field, so new things come about all the time. That's why we need to have a flexible and adaptive perception of it, moving forward. Otherwise, we risk becoming obsolete and unable to practice data science effectively and truthfully.

Machine Learning and AI

As mentioned previously, Machine Learning is related to AI, even if they are not the same thing. First of all, AI is all about emulating intelligence using computers. Intelligence has various aspects ranging from the more high-level (abstract reasoning) to the most mundane (figuring out how to move one's legs to walk). All this is under the umbrella of AI, which is why it is closely linked to robotics. One of the aspects of intelligence, as any psychology professional would tell you, has to do with learning. This is where Machine Learning comes in, even if it exists regardless of the AI field. However, conceptually it makes sense to view it as part of AI since it is related to the learning aspect of intelligence, and people want to keep things simple, even if this means oversimplifying them.

Another differentiator between the two fields is that Machine Learning involves many data-driven algorithms, ranging from Graphs (particularly Trees) to Transductive models, which are models based on a similarity or distance metric. AI models that are used in data analytics, on the other hand, are generally Graph-based, namely Artificial Neural Networks (ANNs). Nevertheless, other specialized AI models make use of alternative processes, such as Fuzzy Logic. These, however, are fairly niche and are not commonly used in data analytics today. Beyond these models, AI also includes optimization systems, particularly those geared toward complex problems. These optimizers can be used in data analytics, too, though their application area goes well beyond that. We talk about these aspects of AI in the book, *AI for Data Science*.

There is also a semantic difference between the two terms, which can be confusing if not considered properly. Namely, Machine Learning usually refers to a model, method, or algorithm in general (a heuristic), while AI usually refers to a complete system. So, when someone deploys an AI for face recognition, for example, this is a complete system that handles most parts of the data science pipeline. A Machine Learning system tends to be much more limited since it usually handles just the data modeling part, which is why it is important to be aware of data engineering as well, since most Machine Learning methods don't include that as part of their process.

Finally, although many Machine Learning models are black boxes, much like the AI ones used in data science (ANNs), they are not all like that. So, if you want to have a transparent model that is not Statistics based, you can do that in both of these fields. Fuzzy Logic systems, for example, are quite transparent and easy to interpret, much like Decision Trees. Note, however, that in most data science applications, especially nowadays, the accuracy of a model is of paramount importance, so black box models are viable options, even if they are tough to understand or interpret comprehensively.

Machine Learning variations

Naturally, Machine Learning involves more than just a couple of methodologies. However, the most important ones, which are also the most commonly used, are the following three:

- Supervised Learning
- Unsupervised Learning
- Reinforcement Learning

Note that at least for the first two types, there are Statistics-based models there too, since these are broader concepts. They are mentioned here as types of Machine Learning for educational purposes. Let's look at each one of them one by one.

Supervised Learning involves models that are trained on some data and then applied on some other, previously unknown data to make predictions. This methodology is crucial for most data science projects and is under the umbrella of predictive analytics. Supervised Learning models in Machine Learning include:

- K Nearest Neighbor (kNN)
- Decision Trees
- Support Vector Machines (SVMs)
- Random Forests (a bunch of different decision trees working together)
- Boosted Trees (a more efficient amalgamation of decision trees)
- Artificial Neural Networks (ANNs), particularly Multi-Level Perceptrons (MLPs)

Note that Supervised Learning is two-fold as it consists of two other methodologies, Classification and Regression, predicting a discreet or continuous variable, respectively. This variable is referred to as a target variable, by the way. All of the models above apply to either one of these two methodologies with some minor modifications. Also, regarding its applicability, it's undeniable as it's responsible for the vast majority of value within a data science system.

On the other side of the spectrum, Unsupervised Learning doesn't involve any variable to predict. This variable may exist in the dataset. However, when applying an unsupervised learning method, we usually don't make use of it in our unsupervised learning model. We may do so afterward, but that's usually at a different stage of the pipeline. Unsupervised Learning involves two methodologies: dimensionality reduction and clustering. Although these are independent of each other, they are often used in tandem since clustering algorithms tend to work better when fewer variables are involved. The latter is a specialty of dimensionality reduction algorithms. Also, we often need to visualize the results of clustering, and reducing the dimensionality of the data is usually a necessary step to enable that.

Dimensionality reduction entails the creation of a new feature set that has fewer features, yet contains a sizable portion of the information in the original feature set. Think of it as a JPEG image created from a Bitmap one or a RAW file as that produced by a D-SLR camera. The latter image is bulky but contains a lot of details, which is why many graphics designers prefer that as their *prima materia*. A JPEG image, however, is much smaller in size and contains enough detail to be useful. The same goes with a feature set that stems from a dimensionality reduction process.

Dimensionality reduction can take place in two ways: feature selection and feature fusion. Feature selection, as the name implies, is all about figuring out which features to keep and jettisoning everything else—it is closely linked to feature evaluation, the process of measuring the information content or predictive potential of a feature or a set of features. Feature selection is particularly useful in cases of a vast number of features, where alternative approaches are impractical. The most common dimensionality reduction strategy involves merging features together to form more powerful information rich features, also known as meta-features. Although we do this through statistical methods such as PCA, ICA, and FA, it is possible to do it through heuristic-based methods too, such as UMAP, which are under the Machine Learning umbrella. We'll look at the most useful dimensionality reduction methods employing Machine Learning in Chapter 8.

As for Clustering, this has to do with finding meaningful groups in the data at hand. These groups are clusters, and their names, usually some integer, correspond to a new variable that the clustering algorithm yields as output. Clustering is essentially a way to develop a labels vector based on the original data. Note that this array of integers may or may not correlate with the target variable, however, since the latter has its information beyond the information related to the geometry of the data at hand. The latter is summarized in the labels variables a clustering method produces, and it can be insightful in and of itself. That's why clustering is often part of exploratory data analysis (EDA), though not as essential as other processes, such as data visualization.

The most notable Machine Learning model related to Unsupervised Learning is the K-means method, which is a clustering algorithm. Not the best such algorithm—it is, however, a good place to start, since it's fairly intuitive and easy to implement from scratch, as an exercise. Here is a good resource for learning more about the ins and outs of K-means: https://bit.ly/3casssl.

Reinforcement Learning is a whole different animal. Namely, it is the methodology involved with providing rewards and punishments to a system as it learns what it is supposed to learn. Usually, reinforcement learning is linked to very specific applications and borrows a lot from AI, due to the complexity of the problems involved. However, it is a Machine Learning methodology that is well-defined and useful. In a nutshell, if the system's output is aligned to the objective, such as a particular classification, it is given some points. In contrast, if it is not aligned, it is not given anything, while in some cases, points may be taken away. Since the system aims to maximize this score function, it gradually learns the patterns we expect it to learn, fairly efficiently. As Reinforcement Learning is fairly specialized, we won't delve into it in this book.

Beyond these three types of Machine Learning, there are a few more, which, however, are too specialized to be of interest to the average Machine Learning learner. Nevertheless, if you are inclined to do so, you can research them yourself. The most important of these Machine Learning types are Self-supervised Learning (a process ideal for limited labeled data), Semi-supervised Learning, and Recommendation Systems.

Machine Learning as a business term

Beyond the technical aspects of Machine Learning, there is also the business dimension of it, something many data science learners tend to forget. However, it is as important as the technical aspect of the craft, while for higher-level positions, it is something crucial.

For a business person, Machine Learning is a means to an end or a business resource. The business people involved in data science, especially the high-level ones, don't care about the methods, the models, and everything else we talked about previously. They care about:

- The effect on the bottom line (ROI).
- How likely it is to be an asset.
- The resources required to make this happen (how much it costs).
- The risk of this whole process—how likely it is to become a liability.

That's it. As long as these requirements are met, it can be a genie that produces the outputs for all they care. Also, according to many business experts, Machine Learning is either part of AI or an umbrella term for everything related to automated processes. In the business world, there isn't a consensus on this yet. That's not a bad thing necessarily, since the people involved in this world at least see the value of this field and its potential. The chances are that they will never understand it the same way a data scientist does, much like the CEO of a car company may be oblivious of the inner workings of the car's engine on the Thermodynamics level. The moment you, as a data scientist, understand and accept this semantic gap between the data science world and the business world, you'll be able to communicate better with project stakeholders.

Naturally, when you are employing a Machine Learning model or a system that makes use of such a model, as part of your data science work, you'll need to explain certain things. It would make sense to avoid details unless specifically asked about them, probably by someone who is versed in our field. It's best to start from the things a business person cares about and gradually make your way to more detailed descriptions of your work. Also, the less technical jargon you use, the better in general. Besides, by using too many Machine Learning buzz words, you may come across as someone who tries to impress.

Also, always make sure you mention the interpretability part of your Machine Learning system since that's something they tend to care about, even if they don't always value it the same way. There is a trade-off between performance in terms of accuracy and speed, and transparency in the system developed in terms of how well you can explain what's happening and which features play the most important role. That's something the stakeholders are aware of, and they would expect you to talk about since it's high-level enough for it to factor in in their decision-making. It's best to assume that this is a requirement unless stated otherwise since everyone tends to want to know why a particular output is what it is, and sometimes how it came about. We'll get back to this topic later on in this book since it's a vast one.

Useful considerations

Machine Learning is complementary to Statistics, and a good data scientist should be comfortable with both frameworks. Besides, there is a trend toward hybrid models these days, something that is beyond the grasp of anyone not comfortable with both data analytics paradigms.

Another thing you need to keep in mind is that Machine Learning is popular today because of AI and that once the hype of the latter subsides, it may revert to being just another technical term. Whatever the case, it may still be useful, even if professionals in this field may use other ways to communicate the value that this field of expertise can bring to an organization. After all, that's what everyone cares about when it comes to Machine Learning: value. Everything else is seasoning, semantics, and marketing.

Finally, Machine Learning innovations take place constantly, so it's paramount that you follow what's happening in this field if you wish to remain relevant. Fortunately, with a tool like Julia, implementing the latest and greatest algorithm in this field is quite feasible and perhaps a rewarding challenge too. After all, Machine Learning is just a state of mind, encapsulating the beauty and efficiency of the data-driven approach to data analytics.

Summary

- It's important to know about Machine Learning since it's an essential part of data analytics today (especially data science). At the same time, it is recognized as a powerful field, even in the business world.

- There are two main paradigms in data analytics: the model-driven approach (conventional Statistics) and the data-driven one (few to no assumptions about the data).

- Machine Learning is under the data-driven paradigm umbrella since its methods rely on the data itself primarily, instead of a mathematical model describing the variables involved.

- Machine Learning is linked to AI as it makes use of some of its methods, but the two fields are distinct.

- Predictive analytics involves methods geared toward making predictions about the data.

- Most Machine Learning methods are related to predictive analytics.

- Machine Learning comprises many methodologies—the most important are supervised learning (predictive analytics), unsupervised learning, and reinforcement learning.

- Although there is a notable overlap between Machine Learning and Artificial Intelligence, they are not the same thing. This overlap involves things like Artificial Neural Network (ANN) models, used for predictive analytics, among other things.

- Business people perceive Machine Learning differently, and there is an inevitable semantic gap between them and the data science professionals. A good data scientist knows how to navigate this and potentially bridge it in data science project communications.

Questions

1. Why is Machine Learning important these days?

2. What's the key difference between kNN and a statistical model?

3. How does Machine Learning relate to AI?

4. Is there a discrepancy between what we as data scientists think of about Machine Learning and what a businessperson may think?

5. Is it better to use a Machine Learning method instead of a model-based one? Explain.

6. Does the Machine Learning know-how from a decade ago apply to today? Why?

CHAPTER 5

Machine Learning libraries in Julia

Although knowing about the Machine Learning packages that are available in Julia is useful in and of itself, it's equally important, if not more important, to know how to use them in practice and understand their functionality in data science projects. Namely, you need to know what kind of data they require as inputs, what outputs they yield, the parameters they have, and the factors in the problem you are solving.

In this chapter, we'll view the most important Machine Learning libraries in Julia, after we introduce the dataset we'll be working with in the next couple of chapters. Following this, we'll start with the libraries essential for data preparation tasks. Next, we'll view libraries related to data models, particularly models geared toward predictive analytics. Afterward, we'll look at various packages related to utilities under the Machine Learning umbrella, as well as some packages that don't fit in any of the categories above, but which are still relevant to data science work.

It's important to pay attention to this chapter as it's instrumental for understanding how the libraries above work, so that you can make use of them in hands-on tasks, as we'll see in the following chapter. Also, make sure you have installed all the packages mentioned in chapter 3 before applying the code in this chapter. To mitigate the risk of any issues, all the packages must be up-to-date, something you can do using the following command (type "]" without pressing enter to enter in the package management mode):

(v1.1) pkg> update

About the dataset

To demonstrate best the use of the various packages used in Machine Learning projects, we first need to have a dataset. For this purpose, we'll make use of the *localization* dataset (http://bit.ly/2lOFOWm), which has to do with the strength of Wi-Fi signals from various devices to four rooms of a house—a situation that probably you can relate to!

The objective of the project related to this dataset is to be able to figure out which room we are in, based on the Wi-Fi signals' strength, given the assumption that there is a single router in the whole house. This dataset is simple, comprising continuous variables used as features and a discreet target variable, namely the room label (in this case represented as an integer). As a result, we'd expect the performance of the predictive analytics model used to be fairly high.

Naturally, the nature of this dataset makes it useful for classification tasks mainly. To make it usable for regression too, we can create a new variable comprising of a subset of the original seven variables and a weights vector w such as [5, 10, 15, 20] and some white noise. This way, the new variable z can be calculated as follows:

$$z = X[\text{indexes of 4 features}] * w + 0.01 * randn(2000)$$

where X is the features matrix, w is the weights vector containing the correct coefficients of the regression model, and randn() is a function for generating pseudo-random numbers to model the noise (these numbers follow the normal distribution). The latter is useful for making the problem more of a challenge for our regression models, so it's not a trivial case. Also, we use 2,000 random numbers since there is a total of 2,000 data points in the dataset. Naturally, we control the effect of the noise with the coefficient of the random numbers array.

Feel free to experiment with different levels of noise on your own. The sandbox section of the Jupyter notebook could be useful for this as well as any other tweaks you'd like to do for this project.

Data preparation processes and libraries

Let's start with some essential data preparation processes and the packages that correspond to them. First, we need to load the *DataFrames* and *CSV* packages, as you can see from the corresponding Jupyter notebook accompanying these chapters. Having done that, let's load the data from the .csv file you download from the dataset's URL, using the following command:

```
df = CSV.read("localization.csv", header = false);
```

Once data is loaded into the data frame *df*, you can handle it just like you'd do if it were in a Pandas data frame. For example, to add a new variable in that data frame that takes six of the seven WiFi signal inputs using the weights mentioned in the previous section, we'd take the following step:

```
df[:RegressionTarget] = Matrix(df[[1, 4, 6, 7]]) * [5, 10, 15, 20] +
    0.01*randn(2000)
```

Note that the Matrix() function makes the whole process quicker since otherwise, we'd need to take each one of the columns of the data frame and multiply it with the corresponding weight, before summing them together and adding the white noise. Although this process wouldn't be excessively cumbersome for a simple problem like this one, it's far more elegant to use the approach shown here.

Before continuing with the Machine Learning packages, let's do some preprocessing of the data using one of the Stats related packages, StatsBase. For example, we could do some normalization of the variables involved:

X = StatsBase.standardize(ZScoreTransform,map(Float64, Matrix(df[1:7])), dims=2)

Note that normalization can take place in two different ways with this function: either using the mean and standard deviation (*ZscoreTransform*) or by making use of the minimum and maximum values (*UnitRangeTransform*). Also, the data used with this function has to be of type "Float" and in the form of a matrix, rather than a data frame. Finally, the standardize method exists in the *MLJ* package, too, so you need to be clear that you are using the *StatsBase* one by including the package name if you have loaded both packages in Julia. Note that when normalization makes use of Z scores, like in this case, it is called standardization. This is also the most common kind of normalization.

Now, we are ready to apply some more interesting methods on the data, something we'll explore in the sections to come. Should you wish to apply other methods, such as the heuristics mentioned in the appendix, you can use this version of the dataset. It's not uncommon to even save the processed dataset into a new data file for easier access to it, particularly if the preprocessing stage is time-consuming, as it's often the case for larger datasets. We'll look into a way to do this toward the end of this chapter.

Libraries for Machine Learning-based data models

Let's continue our exploration of the Machine Learning packages by focusing on those related to data models. Here we'll view the key commands to apply these models. For a more detailed view of their functionality, you can refer to the

Jupyter notebook as well as chapter 6. Also, if you are bold enough, you can check out the corresponding documentation of these packages.

Clustering methods

We'll start with a clustering model, as these are fairly simpler and tend to pop up earlier in the data science pipeline, particularly in the data exploration stage. Namely, we'll look at two clustering algorithms, K-means, and its fuzzy logic variant C-means. In both of them, you need to have the number of clusters predefined (a parameter that takes integer values larger than 1). In this case, we'll opt for four clusters, which is also the most meaningful option considering that the data relates to readings from four different rooms. To adhere to good programming practices, we set this value to a variable (nc) to use in our experiments.

You can use K-means as follows:

```
R = kmeans(XX, nc; maxiter=200, display=:iter)
```

where *XX* is the adjoint matrix of the feature set (something that you can create easily as you can see in the corresponding notebook), and *nc* is the number of clusters, in this case, four. Beyond these fundamental inputs, there's also a couple of additional parameters, namely *maxiter,* which is the maximum number of iterations (something that has to do with the inner workings of the K-means algorithm), and *display* which is what the method displays on the screen as the algorithm runs. Note that only the first two arguments of this function are essential.

As for C-means, the usage is quite similar in this package:

```
R = fuzzy_cmeans(XX, nc, 2, maxiter=200, display=:iter)
```

Note that the parameter after the number of clusters is the fuzziness factor, taking the value of two in this case, which is also the default. This parameter needs to be greater than one for the algorithm to work properly. The default value is fine for this particular dataset, and it's a good starting point in general. Feel free to experiment with other values of the fuzziness factor and observe how it influences the clustering process.

Classification methods

Let's now take a look at a couple of more interesting Machine Learning models, namely classification and regression. We'll begin with classification as it's more suitable for the problem at hand. Namely, we'll apply three models that are often used in data science: a decision tree, a random forest, and another tree-based ensemble method.

For the decision tree case, we can create an instance of this model using the following command:

```
tree = DecisionTreeClassifier(max_depth=3)
```

The only parameter we set here is *max_depth*, which is also the most important one. As the name suggests, this involves how many splits there are going to be in the decision tree at most.

To train the tree, we just need to feed it some training data as follows:

```
DecisionTree.fit!(tree, XX[train,:], y1[train])
```

Note that here *XX* is the features matrix formatted as Float numbers. Also, the features data doesn't need to be normalized—though if it were normalized, it wouldn't affect the result negatively. Finally, *y1* is the target variable for our classification problem, and *tree* is the decision tree model we initiated in the previous step.

Once the decision tree is trained there is some meta-data that is being printed on the screen, yet if we wish to view it in more depth we can use the following command:

```
print_tree(tree)
```

To apply the decision tree, we need to give it some new feature data and use the following code:

```
DecisionTree.predict(tree, XX[test,:])
```

Note that the results are in the form of strings since decision trees in Julia require the target variable to be in this format.

Random forests work in a very similar manner, though they make use of a bunch of decision trees or varying inputs. The code below is how you would create, train, and apply such a model:

```
forest1 = build_forest(y1[train], XX[train,:], n_subfeatures, n_trees, pst,
        max_depth); # build and train model

yhat = apply_forest(forest1, XX[test,:]); # predictions

scores = apply_forest_proba(forest1, XX[test,:], class_vector) # confidence
```

Note that the last part is not essential, but it can be quite useful. It involves calculating the confidence scores of the various predictions made (the result is a matrix having four columns, one for each class).

Finally, AdaBoost Tree Stubs, which are low-depth decision trees in an ensemble setting using AdaBoost, work similarly. Namely, you can use the following code to perform the same function as in the previous case:

```
model, coeffs = build_adaboost_stumps(y1[train], XX[train,:], ni); # build and
    train model

yhat = apply_adaboost_stumps(model, coeffs, XX[test,:]); # predictions

apply_adaboost_stumps_proba(model, coeffs, XX[test,:], class_vector)) #
    confidence
```

Just to clarify, the class_vector parameter is an array containing all the class names, as strings. In this case, it is equivalent to ["1", "2", "3", "4"]. The reason why it's a strings array is that the class labels can be anything, not just integers, as in this case.

Regression methods

Let's now view a couple of examples of regression models using the same packages. Naturally, we'll utilize the new variable we created as a target variable in this case ($y2$). As for the models, we'll use the random forest one as it's generally more popular, at least as a starting point. We won't look into any other models since the process is very similar to that of classification. So, to build and train a regression model using a random forest method, we need to use the following code:

```
forest2 = build_forest(y2[train], XX[train,:], n_subfeatures, n_trees, pst,
    max_depth); # build and train model
```

Notice that the only difference here, apart from the name of the model, is the target variable. Once the target variable $y2$ is analyzed, it will become clear to the model that it's a regression problem. As for applying the trained random forest regression model, the code is almost identical with the one in classification:

```
apply_forest(forest2, XX[test,:])
```

Naturally, there is no method for calculating the confidence of these predictions, since regression systems lack this characteristic in general, something evident through the corresponding models in other programming languages too. This doesn't have to be the case, however, since it is possible to have a confidence score in regression models too. Such a task is not only beyond the scope of this book, however, but also beyond the canon of data science overall.

Machine Learning utilities libraries

Packages related to Machine Learning utilities are quite important, particularly if you want to go into more depth in the processing of data through Machine Learning methods. The utilities we'll examine are two-fold: the ones related to handling labels (for classification problems) and the ones that have to do with model performance.

Performance evaluation functions

We have already looked at some performance evaluation metrics in the Jupyter notebook, but it's good to learn more about the ones on the MLBase package too. For starters, we have the accuracy rate or true positive rate:

```
correctrate(gt, pred)
```

Note that *gt* is the actual labels of the data we are testing the model on (aka Ground Truth), and *pred* is the predicted values that the model yields. Also, both of these are in the form of a numeric vector, so some preprocessing is often required before we can apply this method. Alternatively, we could go with a simpler piece of code that works with any data format:

```
sum(yhat .== y1[test]) / length(test)
```

The confusion matrix is almost as simple as the accuracy rate function, but it also requires the number of classes in the original targets vector (k parameter):

```
confusmat(k, gt, pred)
```

Note that this k parameter has nothing to do with the k of K-means we saw previously. K in the confusion matrix is defined by the dataset in a deterministic way, while K in K-means is a user input, and we may try various values for it before we decide on one to use in our final model.

The Receiver Operator Characteristic (ROC) data is also quite useful when evaluating a classifier, something you can do as follows:

```
ROC = roc(gt, scores)
```

where scores are the predictions before they are made into discreet values and assigned to predefined class labels. In other words, they are the outputs of the classifier right before a decision is made to put a data point in a particular class. The scores variable is closely linked to the confidence vector for each one of the predictions, and is easy to calculate based on the *scores* matrix:

```
conf = maximum(scores, dims=2)[:];
```

Note that you can perform ROC analysis even without the confidence score, using merely the predictions vector. However, the ROC analysis using the scores vector is more accurate overall. Also, the ROC variable that contains the result of this analysis is a composite array that contains the various performance Statistics for each one of the various possibilities examined in the ROC analysis, all of which are used to plot a zig-zag line known as the ROC curve. So, you can select a single point in the ROC curve by indexing it:

r = ROC[50] # take the 50th data point of the curve

We can calculate an additional metric for the classifier, *for that particular instance,* as follows:

precision(r)

recall(r)

f1score(r)

Note that you can get a more general view of the model's performance through the K-fold cross-validation method:

nfoldCV_forest(y1, XX, n_folds, n_subfeatures)

Also, here we use the whole dataset as input since the method does its splitting into training a testing *n_folds* times. The same method is used for performing this evaluation process for regression models, although the metrics used then are different. We'll talk more about K-fold cross-validation (KFCV) in the next section. However, it's good to remember that model-specific KFCV methods like this one are generally preferable.

Although the ROC analysis approach is an excellent way to evaluate a binary classification model, the method here is not ideal if you wish to plot the whole curve. However, you can do a full ROC analysis, including a plot, using the *ROC* package. We'll look into this in Chapter 7, where we'll tackle a binary classification problem.

Cross-validation functions

Cross-validation functions come in two varieties, the ones for splitting the data and the ones for applying some metric on the different splits. Since the latter are underdeveloped and counter-intuitive, we'll focus on the former kind here.

Besides, once you've got the data split up, it's not difficult to do your cross-validation afterward using the evaluation metrics we saw in the previous section. First of all, we have a K-fold sampling method:

Kfold(n, k)

where n is the total number of data points, and k is the number of folds (usually something between three and ten, though there is no restriction for the values of this parameter). Again, this parameter is something we define, having nothing to do with the number of classes in the dataset or the number of clusters.

As for the Leave-One-Out method, which is useful for particularly small datasets, the corresponding command is:

LOOCV(n)

where n is the total number of data points, like before.

Finally, the Random Subsampling method:

RandomSub(n, sn, k)

where sn is the number of samples to be created.

Note that all of these methods perform some kind of partitioning of the dataset, be it in k partitions, n partitions, or sn samples; they don't perform any evaluation of the data whatsoever. However, given what we have seen so far, it shouldn't be difficult to do that ourselves.

Label related functions

Labels are very important, and they often require some data engineering work of their own, particularly when it comes to classification problems. For example, some models prefer the labels to be in the form of a binary matrix, with each column corresponding to each class. What's more, they yield results in the same format, so the aforementioned transformations are useful in more than one way.

First of all, we have the label encoding function that can help map labels into integers (or numbers in general), something that can be useful in some models, as we saw previously:

```
labelenc(y1)
```

Naturally, there are different ways to encode the labels vector. One very common way is through zero-one encoding, something particularly useful when applying a binary classifier. This encoding involves two values, 0 and 1, as numeric values, populating the labels vector. So, we can check to see if the labels follow such a pattern through the following command:

```
islabelenc(y1, LabelEnc.ZeroOne)
```

A quite useful method for transforming labels into a different set that is perhaps more meaningful is:

```
convertlabel(new_labels, y1)
```

What this particular command does is convert all the values of *y1* to those in the vector *new_labels*, which can be something like this:

```
[:Room1,:Room2,:Room3,:Room4]
```

Naturally, you don't have to use symbol values to populate this vector, as sometimes integers can do the trick:

```
convertlabel(LabelEnc.Indices{Int}, y1)
```

Some more sophisticated encoding options can involve binary matrices, having one row for each class, and one column for each data point. Having a matrix as a label is sometimes essential for the more advanced classification models. Here is how you can do this:

```
convertlabel(LabelEnc.OneOfK{Bool}, y1)
```

Of course, you can also do the same vector to matrix transformation using float numbers:

```
convertlabel(LabelEnc.OneOfK{Float64}, y1)
```

An interesting application of this is through the classify() function of this package, that can perform the prediction based on the confidence scores for the various classes (scores matrix):

```
MLLabelUtils.classify(scores', LabelEnc.OneOfK)
```

Note that the scores matrix needs to be of the same orientation as the OneOfK encoding option, whereby the classes are as rows instead of columns. That's why we need to use the single quote operator ('), which flips a matrix so that the rows become columns and vice versa. Also, since the classify function is bound to exist in another one of the packages we use when dealing with predictive models, it's essential to specify the function's package, which is the MLLabelUtils package.

Additional Machine Learning libraries

Beyond the above packages related to Machine Learning processes, there are a few more that don't fit in the previous categories but are still good to know as they facilitate data science work immensely. Namely, the distances, TSne,

Gadfly, and JLD2 packages, all of which can be quite useful in various Machine Learning projects.

Distances library

The distances library is a collection of various distance metrics implemented as independent functions. Distance metrics may sometimes be useful, particularly for heuristics, such as when the similarity between two features or between a feature and the target variable is required. After all, almost all similarity metrics are a linear transformation of a distance metric. However, a non-linear transformation can also be used, such as 1 / d, where d is the distance value. Note that this last transformation is a bit crude and may not be so practical in cases where distances are very small or even zero.

Also, even though distances can be easily coded without requiring specialized knowledge of mathematics, this package has optimized versions of the various distances metrics. These versions are faster than the simple distance methods someone could put together using loops, even though loops are quite fast in Julia. To apply a particular distance metric *dist* on two vectors, *x* and *y*, you need to use the following code:

```
Distances.evaluate(dist, x, y))
```

or

```
dist(x, y)
```

Note that each distance metric is in a peculiar form, with each one ending in a pair of brackets with nothing in them. You need to refer to each metric using that form. For example, when applying the Cityblock distance metric, we use:

```
Cityblock()(x, y)
```

If you wish to apply a distance metric *dist* to a bunch of vectors that are part of a matrix X, either are rows or columns, you need to use the code:

```
pairwise(dist, X, dims=d)
```

d is the dimension of the matrix you wish to use (1 = rows, 2 = columns).

TSne library

The TSne library is one of the most established Machine Learning packages, and one that has been translated into many programming languages. It's particularly useful when you want to plot something on a two-dimensional plane to visualize it more easily for a more in-depth EDA.

To apply TSne, you need to rescale the data to ensure there are no extremely high distances among the data points, using a custom re-scaling function such as the one provided in the TSne demo:

```
rescale(A; dims=1) = (A .- mean(A, dims=dims)) ./ max.(std(A, dims=dims),
    eps())
```

The eps() part of this function is a very small number, which has to do with the computer's accuracy level when it comes to variables. The max(…, eps()) part of the function ensures that even if the standard deviation of A is zero, this part will still take a non-zero value, namely the eps() one. For the computer used in this book, eps() is approximately equal to 2e-16.

Note that the default dimension for re-scaling is 1 (rows), meaning that the method interprets the rows as data points—the columns of the dataset A are the features. To apply this re-scaling method, type:

```
RX = rescale(X, dims=1)
```

Afterward, you can apply TSne on the re-scaled dataset *RX* as follows:

```
tsne(RX, nd, p, s, epsilon)
```

where *nd* is the number of dimension at the output (usually takes the value 2 as it's easier to plot 2-D data), *p* is the perplexity parameter (values between 5 and 50) denoting how much gravity to put on local variations (low perplexity values), *s* is the number of steps before the method terminates (default value is 1000), and *epsilon* is the learning rate, which is how fast changes takes place in the algorithm (values between 1 and 20).

Note that T-SNE is a stochastic algorithm, so the results every time you run are bound to be somewhat different. Also, although it does a good job of reducing the dimensionality for easier plotting, the resulting dataset may not reflect the geometry of the original one properly. Finally, T-SNE has been around long enough to be trustworthy. However, newer dimensionality reduction methods like UMAP appear to be more promising and generally more useful. As a result, it is recommended you don't spend too much time on this library, particularly if you wish to apply dimensionality reduction to larger datasets (either in terms of data points or in terms of features).

Gadfly library

Since plotting is very important in EDA, it wouldn't be right to neglect delving into the Gadfly package, which is one of the best ones in Julia. Other visualization packages are also good, but Gadfly is easier to use and better documented than anything else in the Julia ecosystem. To create a basic plot between two variables, you need to apply the following code:

```
plot(dataframe, x = :variable1, y = :variable2)
```

So, if we were to apply this to the first feature and the RegressionTarget variable, we'll have something like the plot in Figure 5.1

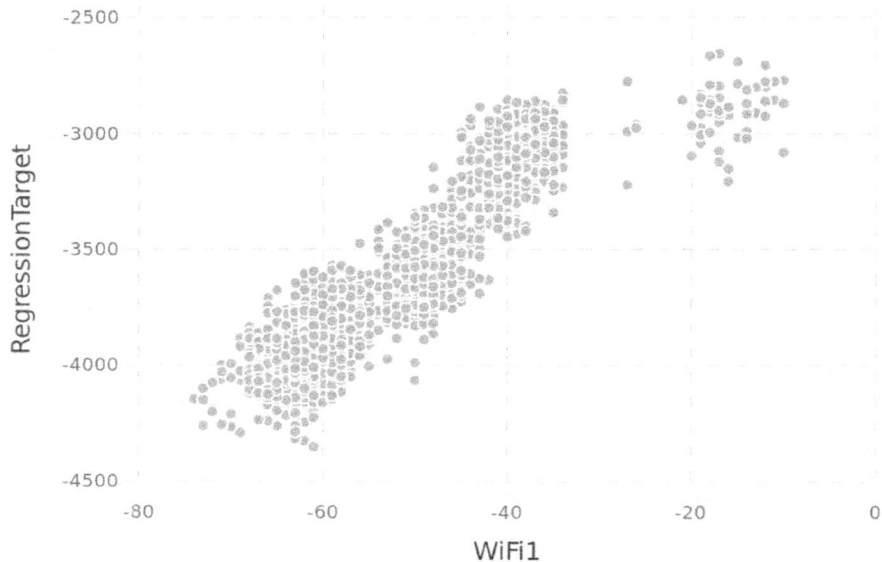

Figure 5.1 The first variable versus the RegressionTarget one.

You can code labels as colors if you like, using the *color* parameter. So, if you want to show how the first two features of the dataset correspond to different rooms, you can use this code:

```
plot(df2, x = :WiFi1, y = :WiFi2, color = :Room)
```

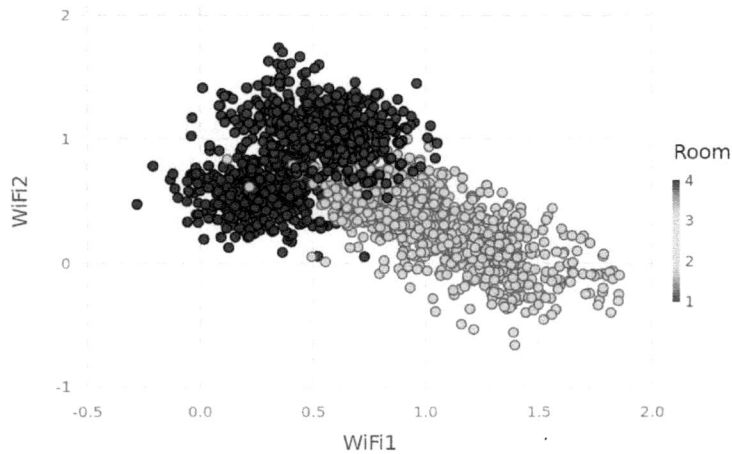

Figure 5.2 The first two features and the *room* variable (color).

Or if you want to see how the T-SNE variables look like in relation to the rooms, you can type:

```
plot(df3, x = :x1, y = :x2, color = :x3)
```

where *df3* is a dataframe containing the two T-SNE variables as well as the *room* one from the original dataset.

Of course, histograms are particularly handy for EDA so it's good to be able to handle those too, something possible through these commands:

```
plot(dataframe, x = :variable_name, Geom.histogram)

plot(dataframe, x = :variable_name, Geom.histogram(bincount=30))

plot(dataframe, x=:variable_name_1, y=:variable_name_2,
     Geom.histogram2d(xbincount=40, ybincount=40)) # a more elaborate
     2-D plot
```

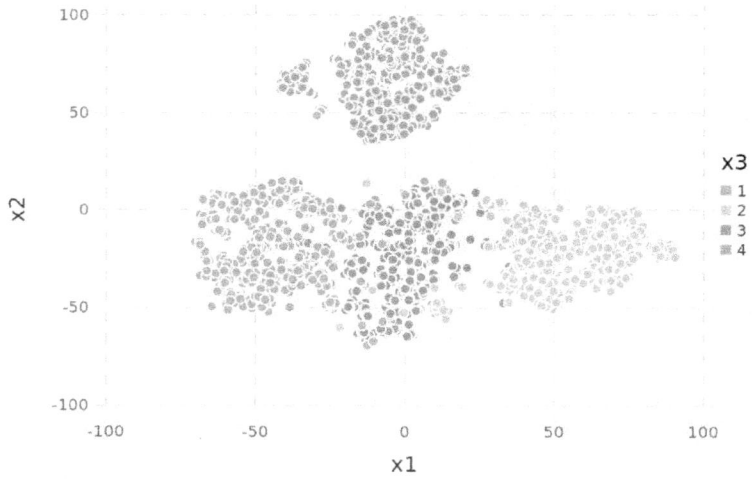

Figure 5.3 The T-SNE variables (*x1* and *x2*) in relation to the *room* variable (color).

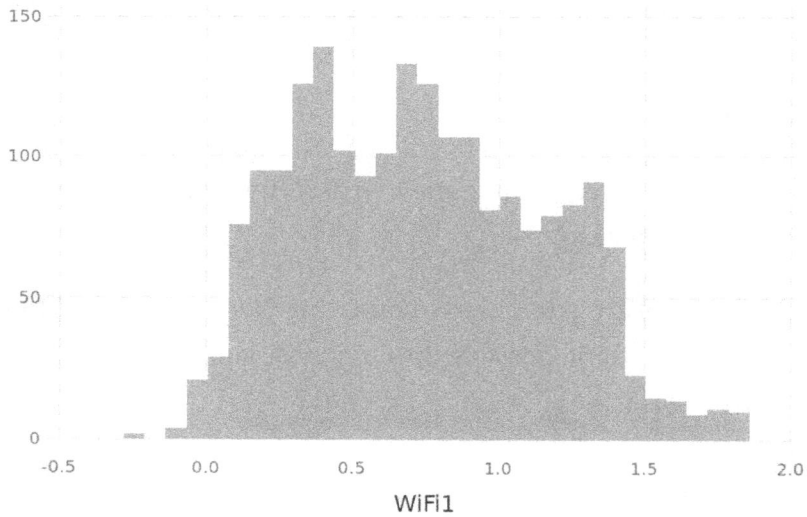

Figure 5.4 Histogram of the WiFi1 feature using 30 bins.

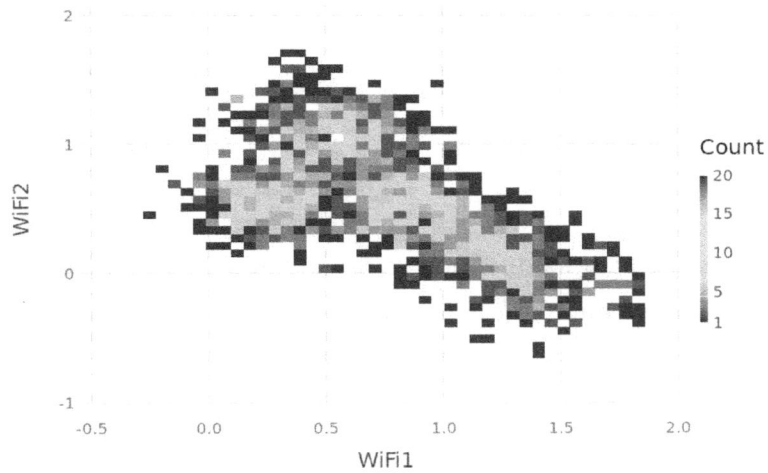

Figure 5.5 Two-dimensional histogram of features WiFi1 and WiFi2.

Finally, as heatmaps are a powerful tool for visualizing matrices and such, it's useful to become familiar with this command:

spy(matrix)

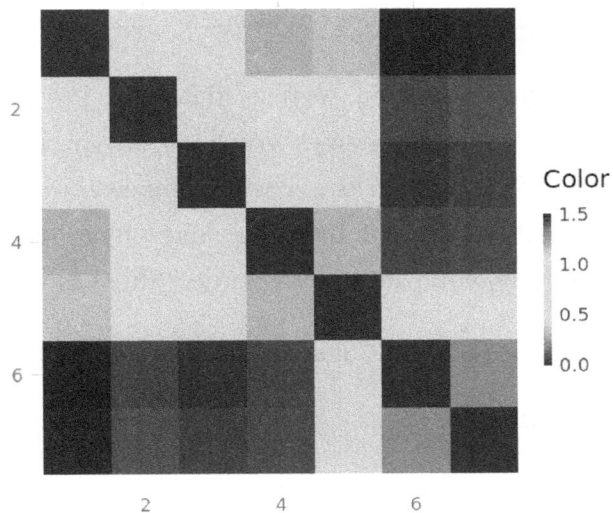

Figure 5.6 Heat map of the matrix of Jaccard distances of the various features.

Beyond these, there are several other plotting options, too many to list here, so feel free to check out the official Gadfly website: http://gadflyjl.org/stable.

JLD2 library

Although Julia is great at loading and saving data in CSV format, it's sometimes more practical to save data in JLD format, which is a native Julia data structure. JLD2 is a newer version of the original JLD one, and it's the one most commonly used today.

To make use of the JLD2 package you need to also use the FileIO package, since the latter is a dependency of the former. Run this command to add the FileIO package:

```
using JLD2, FileIO
```

To save with JLD2, you need to write the following command:

```
save(filename, Dict(varname1 => var1, varname2 => var2, ...));
```

where *filename* is a string variable such as data.jld2, just like *varname1* and *varname2*. The latter two are the names you refer to variables var1 and var2 respectively, in a dictionary structure you create, enabling storing different variables in the same file, even if these variables have completely different dimensionality. Naturally, you can save as many of them as you like.

To retrieve data using the JLD2 package you need to use the following code:

```
alldata = load(filename);
```

and then retrieve the specific variables from the dictionary variable *alldata*, writing something like:

```
X = alldata[varname1]
```

Note that you don't need to use the same variable name to store the data in the *alldata* dictionary. As long as you are clear as to what you are accessing, you can name the variables whatever you like. If you are not sure about the variable names, you can always get that information from the dictionary itself:

```
keys(alldata)
```

Although the files JLD2 creates tend to be bulky, they are practical, particularly when you use them to store a model or any complex data structure (e.g. a data model's parameters and any relevant metadata). Unlike CSVs, which require the data to be in a table-like format, dictionaries can store all kinds of objects in them. The same paradigm is used nowadays in various databases, such as MongoDB, due to its flexibility.

Useful considerations

Although Machine Learning packages are fairly straightforward in Julia (as well as any other high-level programming language), there are some things you need to keep in mind. First of all, the available packages can reflect only the Machine Learning know-how that has been tried and tested by a large enough set of users. State-of-the-art Machine Learning methods may still be unavailable through any packages, particularly methods that employ unconventional approaches and proprietary know-how. In other words, if you want to have an edge in performance using the freely available packages, you need to have excellent data at your disposal. Perhaps that's why companies and other organizations are very protective of the data they have and are reluctant to share it with others, even if it isn't private or if it's been anonymized.

No matter how advanced the methods, we still need a sufficient amount of the right data. The methods themselves can only get you so far; to go further, you'll need to have something that your competitors don't: valuable information—something that usually accompanies the right datasets. Of course, some of these datasets may not be as easily accessible as we would like, but that's expected since quality data usually comes with a price tag. This is particularly the case when it comes to labeled data, something that often requires a human curator to complete this task.

What's more, new Machine Learning packages come about regularly, so it's good to be on the look-out for them. The best packages are yet to come, even if they may not become super popular any time soon. Just like true randomness without the use of an external device is not common knowledge, the truly great Machine Learning systems may not be something found in the Julia ecosystem as they are bound to be proprietary.

Furthermore, as useful as the Julia Machine Learning packages are, it's equally important to use them properly, to make the most of them for your data science work. If you are limited in your understanding of the algorithms involved or how their hyper-parameters are set, don't expect wonders in performance or spectacular results. They are a useful tool but not something as simple as a Swiss army knife! Perhaps that's why there are so many of them complementing each other in various ways. Our suggestion is to use your custom-made code when it comes to Machine Learning algorithms, particularly more advanced ones, since the learning curve of these packages, coupled with the ease of prototyping in Julia, makes this alternative a viable option.

Finally, even if Machine Learning packages in Julia are quite developed nowadays, it's important to also pay heed to other data-related packages. After all, not all data analytics tasks are linked to Machine Learning. Sometimes a different approach, such as Statistics-related, is needed, while other times, it's all about cleaning and formatting the data properly that matters (data engineering). Machine Learning is a powerful tool in data analytics work, but it's definitely not the only tool and not the one most suitable for every project.

Summary

- Learning about and familiarizing oneself with the Machine Learning libraries in Julia is crucial for making use of them in data science projects.

- Before using the libraries mentioned in chapter 3, we need to make sure that they are all installed and up-to-date.

- It's useful to have a dataset that's not too complex to try out and explore the methods of any data analytics package in Julia, such as the ones related to Machine Learning tasks.

- The first step to using any of the Machine Learning models available in Julia is to get the data loaded (usually into a data frame) and prepared (cleaned and normalized).

- The *standardize* function from the *StatsBase* package takes care of the normalization of a dataset, whether it is using the mean and standard deviation method or the min-max approach. These functions don't change the nature of the data, merely its scale and descriptive Statistics.

- For clustering tasks, the *kmeans* and *fuzzy_cmeans* methods from the *clustering* package are both good options. However, there are a few more clustering algorithms available on that package to explore, even if they are not as popular overall.

- Regarding classification or regression tasks, there are several packages, such as *DecisionTree* and *MLJ*. The necessary functions for these are *DecisionTree.fit!()* for training a decision tree, and *DecisionTree.predict()* for applying a decision tree.

- The *MLBase* and *MLLabelUtils* packages have several useful ML utilities. In MLBase, there are methods such as the confusion matrix (*confusmat()* function) and the ROC analysis (*roc()* function), while in the MLLabelsUtils package there are various methods for changing the labels encoding (*islabelenc()* and *convertlabel()* functions) as well as one for performing classification based on the confidence scores (*classify()* function).

- Other useful methods for ML work are from the *Distances, TSne, Gadfly,* and *JLD2* packages. Namely, *Distances* has various distance-related functions (for similarity assessment), TSne involves a method for reducing a dataset to a dimension that enables easy plotting, Gadfly includes a variety of visualization options such as scatter plots, histograms, and heat maps, and JLD2 involves the storage of all kinds of data in a flexible data structure.

Questions

1. How can you make sure that a given package X is up-to-date so that you can use it for your Machine Learning task?

2. What's the difference between a data frame and a matrix?

3. How would you use the CSV package to handle a data file that contains headers? What happens to the header metadata?

4. Why would you want to normalize the data at hand? Are there other ways to do so apart from the ones mentioned in this chapter?

5. Can you combine clustering and classification for better results? What about Can clustering and regression?

6. What's the most important Machine Learning utility introduced in this chapter?

7. Are there any assumptions you need to make about the data in order to use the methods introduced in this chapter?

8. Can you work a data science project using solely Machine Learning packages? Explain.

CHAPTER 6

A Data Science Project Using Machine Learning

Let's now look at a couple of examples in more depth, so that you get a better understanding of how the material mentioned in the previous chapter works in practice. After all, knowing a few specific functions may be useful, but it's even more useful when you know how to weave a narrative around the code and create a coherent project. Similar to the previous chapter, this one is accompanied by a Jupyter notebook that provides the script for this narrative (both figuratively and literally).

In this chapter, we'll examine a particular dataset that is reasonably simple: Wine Quality (https://bit.ly/2Vf9M5k). We'll analyze that data as a regression problem as well as a classification one. Afterward, we'll make technical and non-technical conclusions from this analysis.

For additional practice, feel free to experiment with this dataset on your own and see if you can improve the models further or build alternatives with better performance. Remember, the Sandbox area of the notebook is your friend.

Dataset overview

The wine quality dataset involves 12 characteristics of about 6,500 red and white wines. Most of the characteristics expressed as variables of the dataset are objective measures of a wine, such as its acidity level, pH, and alcohol level, while the wine's quality is subjective, as gauged by experts, on a scale of 0 to 10.

The wine's quality is also the variable we are trying to predict, using all the other ones as inputs. So, the main question of the project would be, "how can we predict the quality of a wine given its objectively measurable characteristics?" The answer will be a predictive model with wine quality as the target variable.

However, what if the question is, "what are the key differences among all these wines, apart from their quality levels?" Such a question could be answered by clustering analysis (a clustering model), by examining the centroids that ensue. Beyond these two questions, there may be other ones you could ask and experiment with on your own (refer to question 5 at the end of this chapter).

Here is a scientific paper based on this dataset, published by Elsevier: https://bit.ly/34IzaDw. However, the wine quality dataset is a benchmark for not just researchers, and has proven to be a very useful dataset for all kinds of predictive analytics systems, particularly regressors. Also, it can be used differently than the original research—a mentee of mine once tackled this dataset using alcohol level as the target variable and got some very interesting results. The fact that the alcohol level variable has a fairly large variance (larger than the wine quality variable) must have helped.

An interesting peculiarity of the dataset is that it involves two main types of wines, red and white, provided as two separate data files. This makes the dataset lend itself for different kinds of analyses, by taking wine type into account. Also, the wine type is highly unbalanced, as there are many more white wines in the dataset than red ones (4,898 versus 1,599, respectively). Also, for privacy reasons, the names of the actual wines are concealed, something fairly common in data science projects. To avoid any unnecessary complexities, in the analysis of this chapter, the wines are going to be examined as a single dataset, comprising both red and white wines. In addition, the color of the wine is not going to be used as a feature since it is bound to be a poor predictor—there are both good and bad wines in both wine colors, something you can deduce yourself with some descriptive statistics on the wine quality variable of the two subsets. Finally, most of the wines have quality levels in the middle values, making exceptionally low-quality and exceptionally high-quality scores harder to predict.

Data exploration through clustering

Let's start by exploring the dataset using clustering to see if there are any underlying patterns. Before we can apply clustering, we need to load the data and normalize its feature set. So we'll need to first normalize the data, either using standardization or the fixed range method (minmax normalization), as distances are involved in the clustering process. Naturally, if there are redundant variables in the dataset, we'll need to jettison them before we commence with the clustering. You can check the Jupyter notebook for the specifics of this, or you can try to do this on your own.

After the data is ready for modeling we can apply the Kmeans clustering method as follows:

```
R = kmeans(X_final', k; maxiter=200)
```

where k is the number of clusters. Naturally, we'll iterate over various values of k before we can figure out the optimal value of this parameter.

Note that to evaluate the clusters kmeans creates, we'll need to calculate a metric like Silhouette. This metric, however, requires the pairwise distances of the data points of the dataset. We can compute the latter using the code:

```
D = pairwise(Euclidean(), X_final, dims=1)
```

We can use other distance metrics too, though it's important to use the same metric in both the distance matrix calculations and the clustering algorithm. The latter uses Euclidean distance as a default.

After this, for each value of k we can calculate the Silhouette metric using this code:

```
mean(Clustering.silhouettes(R, D))
```

Note that we calculate the mean of the silhouette scores since individual Silhouette scores are not that useful for evaluating the clustering result.

Naturally, we'll work with the value of k that maximizes the Silhouette metric. In this case, this is 2, yielding a Silhouette score of around 0.26. This score is not particularly impressive as Silhouette can go as high as 1.0, but it's not too bad either. If it were negative, that would be a problem.

As for the centroids, let's see what they look like using the corresponding command:

```
show((R.centers)')
```

[**0.784906** 1.15619 -0.344739 -0.591479 0.910328 -0.812097 *0.53596* 0.816235
    -0.0875035; **-0.281496** -0.414653 0.123636 0.212126 -0.326477
    0.291248 *-0.192215* -0.292732 0.0313819]

From these values representing the centroids in the 9-dimensional space we can see that the largest differences are in the features *fixed_acidity*, *volatile_acidity*, and *sulfates* (depicted above in underlined, bold, and italics respectively), meaning that these three factors are the most important for discerning the two clusters.

The fact that there are two clusters also shows that if we were to split the target variable, good and bad wines would be more meaningful. However, this is not an absolute rule as it's more a rule-of-thumb.

Predicting wine quality

Regression

Let's now take a stab at this problem the way it was designed to be tackled, namely through regression. We'll focus on the random forest method as it's a generally good model for benchmarking. To check if a particular regression system is any good, we can do K-fold cross-validation on it, using this code:

```
nfoldCV_forest(y, X_final, n_folds, n_subfeatures)
```

Note that the *n_subfeatures* parameter is the number of features to use in each tree. The value of 3 seems to work well enough. Feel free to experiment with other values of this parameter as well as other parameters of the random forest regressor, such as number of trees. By "well enough" we mean that the mean square error (MSE) and the other evaluation metrics the *nfoldCV_forest* method yields are fairly stable—their fluctuation among the folds is quite small. The coefficient of determination (R^2) in particular is relatively good, but it could be better, something that makes exploring the classification approach to this problem a worthwhile alternative.

Should we wish, we can then build a new random forest regressor and test it on a different sample, using the code that follows:

```
forest = build_forest(y[train], X_final[train,:], n_subfeatures);
```

```
yhat = apply_forest(forest, X_final[test,:])
```

Classification

Since regression may not figure out what's a good wine, or perhaps our taste buds are not sensitive enough to tell the difference between a wine of quality level 7 and quality level 9, classification is also a worthwhile option for this problem.

There are various ways to perform classification for this problem. Since there aren't that many unique values for the quality variable, we could even use the data as is. In practice, however, we usually define a threshold that splits the data into two distinct groups. These are used as the classes of the dataset and that (binary) variable containing them is our new target variable. In this case, we use the mean of the quality scores as that threshold and create our new output variable as follows:

```
m = mean(y);
yy = Int64.(y .>= m);
```

Due to the nature of the y variable, the *yy* binary target is going to be unbalanced, which is unavoidable due to the nature of the data, even with a different threshold. Alternatively, it could be that the wine quality variable is split in three, making this a multi-class classification problem, instead of a binary one. That's something you may want to explore afterwards, for additional practice.

For the classification experiment, we'll look into two models, a simple decision tree and a random forest so that we get a better sense of perspective. Also, individual decision trees tend to be better in classification, which is why we refrained from using them in the regression experiments previously. We start by training and testing a single decision tree using 5-fold cross-validation:

```
tree = DecisionTreeClassifier(max_depth=3);

accuracy = cross_val_score(tree, X_final, yy, cv=5)
```

The performance of the classifier based on the average accuracy score is underwhelming, both fairly low and unstable. However, with a different maximum depth such as 7, we get a somewhat different performance. Upon closer inspection of the values in the various folds, it seems that this performance is still not reliable enough. The reason for this is that in one of the folds, the accuracy was much lower than the others. This unstable behavior of the classifier spells out trouble, if left unattended.

The random forest classifier yields somewhat more promising results:

```
accuracy = nfoldCV_forest(yy, X_final, n_folds, n_subfeatures)
```

We can try different configurations by changing the number of subfeatures, for example. However, for this data, the best performance is yielded when this parameter takes the value of 3. Naturally, this predictive analytics system lends itself to further optimization since there are other parameters to tweak, such as the number of trees and the maximum depth of each tree.

Conclusions and next steps

Technical conclusions

We would normally exhibit our conclusions in a presentation or report. This section summarizes the conclusions—something that normally appears inside the Jupyter notebook where the analysis takes place.

First of all, some of the features of the dataset are correlated. Namely, free_sulfur_dioxide and total_sulfur_dioxide as well as density and alcohol (features 6, 7, 8, and 11, respectively). To mitigate the chances of over-fitting and to reduce potential computational overhead, one feature from each one of these pairs is removed. We chose to keep the features that are most correlated with the quality variable (target). After all, correlation with the target variable is a good proxy for a feature being a good predictor in a regression problem.

As regards to cluster analysis, this process revealed that there are two distinct groups of wines; the first one comprises the wines of low fixed acidity, low volatile acidity, and low sulfates level, and the second group includes all those wines having high fixed acidity and alcohol level. Note that in both of these groups, the other characteristics vary too, but not as much as to constitute discerning factors.

When it comes to predicting the quality of the wines based on a regression model, the results are good but not overwhelming. A clear-cut mapping between the characteristics of a wine and its quality (based on an expert's view) does not exist, at least not based on the available data. However, we can predict the quality with some accuracy (coefficient of determination = 0.69 approximately).

Predicting the quality of the wines based on a classification model is more promising. By using the mean of the quality level as a threshold (below which a wine is considered bad and above which good), the target variable turns into a binary one (0 = bad wine, 1 = good wine). The overall accuracy rate is quite good (mean accuracy = 97.5%) and stable (not much fluctuation in the 5-fold cross-validation setting).

In both the regression and the classification cases, the random forest with three sub-features seems to perform best, as it is an ensemble model. However, additional parameter tweaking might yield even better results.

Not-so-technical conclusions

What if some of the project stakeholders stop listening to you as soon as the talk gets technical? Well, that's where the non-technical conclusions come in handy, plus they are a great segway to the big picture conversation that is bound to ensue.

A not-so-technical conclusion is that we can save the organization resources by bypassing the wine expert consultant needed to evaluate a new wine. After all, if we have a reliable enough mapping that's also quite accurate, we can estimate the quality of a wine on our own. A diplomatic option might sound like this: "we can mitigate the usage of wine experts for the more expensive wines, thereby saving money while at the same time making the most of their expertise."

Another not-so-technical conclusion we can arrive at is that there are two broad kinds of wines, given their objectively measurable characteristics—not taking into account the quality variable. Since we don't know much about these groups, we may explore them further and see what kind of relationships they may have with the quality variable, as well as the binary quality level (good-bad wines). This can take the form of a new iteration of this data science project.

Additionally, we can say that beyond these objective factors of the wines, there may be other pieces of information that are worth looking into as they play an important role when it comes to selecting a wine. For example, the price of the various wines, as well as their country of origin, could be relevant factors to consider. Once additional data is acquired to enrich the existing dataset, we may explore this matter anew, going deeper on it, something that can also take place in the new iteration of this project.

What's more, there are limitations to our analysis. Our sample was fairly small and contained two broad types of wines (red and white). However, there are other types worth considering, such as rosé and green, something that may skew the results of this analysis (green wine is actually one of the types of wine the Portuguese take pride in, for good reason). Also, the very high and very low qualities were fairly under-represented, so accuracy in these parts of the quality spectrum may be lacking. Finally, the models tested here were relatively few, as this was a proof-of-concept project. Other, more sophisticated models are bound to yield better results, something that may be worth considering before deploying the model.

Useful considerations

Beyond these examples, there are several other ones you could (and should) try out for additional practice. Unlike real-world projects, this case study didn't require a lot of data engineering. What's more, hands-on Machine Learning projects tend to illustrate a particular set of methods. In real life, you'd try lots of different strategies, including combinations of these methods. In addition,

Machine Learning projects are not just code and headings in a Jupyter notebook. If you want to share them with other people, particularly people who are not versed in data science, it's best to provide some explanatory text, too, particularly when it comes to the conclusions you draw from all of your work on that project. Otherwise, it may seem confusing and even unauthentic, since you could have copied and pasted that code from elsewhere. This is something that you can apply in every data science project taking the form of a Jupyter notebook, regardless of the programming language you use.

Moreover, hands-on Machine Learning projects are the best way to learn data science and related tools. Remember that there is always more depth to whatever you know on this subject, and as technologies evolve, it's always useful to revise known methodologies with new or refurbished tools. Perhaps that's why data science is a field that never gets old, at least when it comes to hands-on work. Even if parts of it seem repetitive at times, there is always something new to try and eventually master—the programming tools involved continue to evolve.

Finally, even if advanced methods are more appealing and capture most people's attention these days, the conventional methods still have value in data science, particularly when it comes to their transparency. Everything has its place, and even the more humble tools have something to offer if you know how to use them.

Questions

1. Does it matter which package you use for a particular Machine Learning model, in Julia?

2. Could you use random sampling over and over again instead of K-fold cross-validation, when evaluating a predictive analytics model?

3. How can the output of a clustering algorithm be evaluated?

4. What's the most important thing in a hands-on Machine Learning project?

5. What other questions could you ask for this particular data science project?

CHAPTER 7

Code a Machine Learning Model from Scratch

Although Julia has many useful Machine Learning packages to facilitate data science work, sometimes what we need is not available in the Julia ecosystem. In these cases, we need to build something from scratch, a quite daunting task for many people. However, due to the high performance the language offers, this task is not as difficult or as time-consuming as it may first seem.

In this chapter, we'll explore how we can code a Machine Learning model from scratch. We'll look at a model that's not mainstream even though it's quite useful, namely the Fuzzy version of the k Nearest Neighbor model (FKNN), developed in 1985 by J. M. Keller and his team. Although this model was developed over three decades ago, it is still being used as a KNN variant. A fairly recent paper on the topic is titled, "A fuzzy KNN-based model for significant wave height prediction in large lakes" (http://bit.ly/2uu1OKp).

Although the FKNN model can be used both for classification and regression, it is mostly used for classification. Since kNN and its variants are better suited for classification, we'll mainly focus on this task, including the design, requirements, implementation, testing, and showcasing of the model. As a bonus, we'll also look at how you can build a basic K-fold cross-validation process from scratch too, as well as another auxiliary function.

Design of a Machine Learning model

Designing a ML model is fairly simple, as it's usually the case that we make use of pre-existing algorithms, such as those made available through a scientific publication. Naturally, most of the latter tend to be cryptic, unless you are well-versed in the terminology and excessive vagueness that often characterize the math and programming aspects. However, given enough time and patience, you can make sense of them and get something usable out of the text in these papers.

A common practice among programmers and people who design information systems is to create a general outline representing the inputs and the outputs of the system, such as this simple UML diagram:

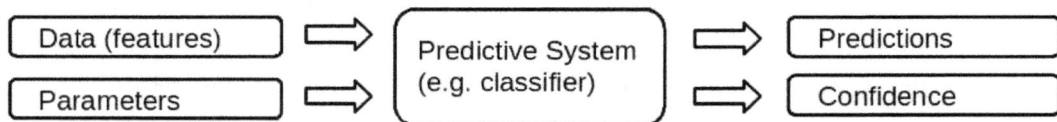

Figure 7.1. Simple UML diagram representing a predictive analytics ML system. Note that the confidence output is not always made available.

Of course, the high-level design of the system in Fig. 7.1 is but the beginning of the whole process and serves more as a guideline and a supplementary document for the more detailed description in Fig. 7.2. The latter oftentimes takes the form of pseudo-code, as it's easier to "translate" pseudo-code into programming code than a flowchart. Before doing so however, let's get a better idea of how all this fits in the broader pipeline by examining the requirements of the ML model.

Start

Input sample x

n: Number of labeled samples

K: Number of nearest neighbors

$\{x_1, x_2, \ldots x_n\}$: A set of n labeled samples

Set K ($1 \le K \le n$)

Initialize $i=1$

Compute distance between x and x_i

$i \le K$?

Yes

Include $u_i(x)$ in the set of K-nearest neighbors

No

Search and find x_i closer to x than any previous nearest neighbor

- Delete the farthest of the K-nearest neighbors,
- Include x_i in the set of K-nearest neighbors.

$i = i+1$

Are K-nearest neighbors to x determined?

No

Yes

Initialize $I = 1$

Compute $u_i(x)$ using equation 3

Are fuzzy membership degrees of all classes determined?

No

$i = i+1$

Yes

End

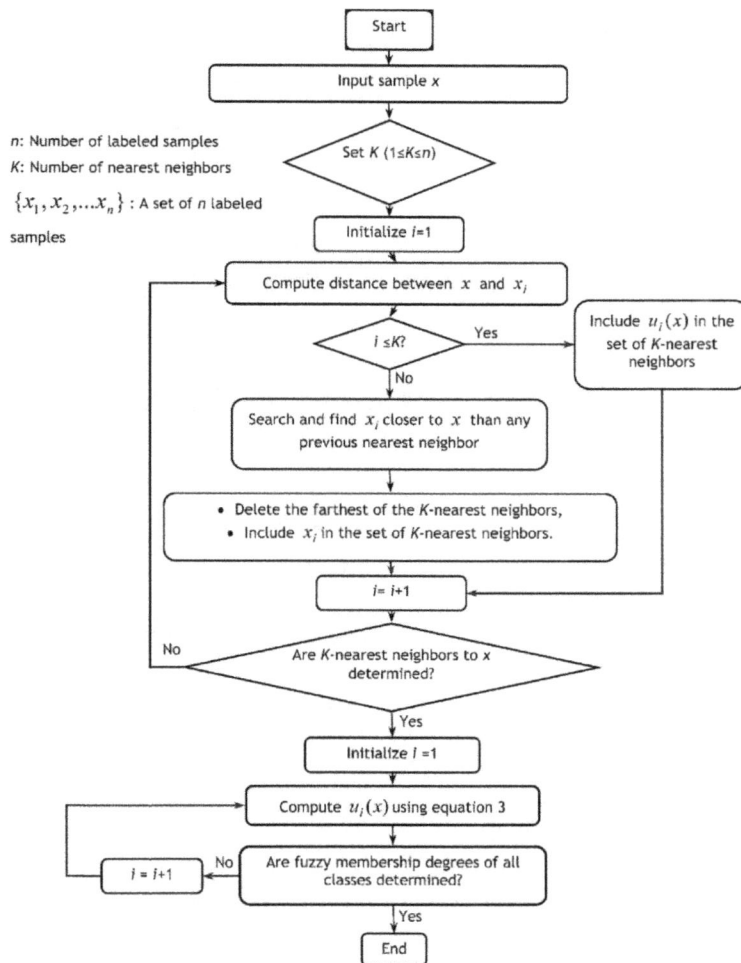

Figure 7.2. Detailed diagram of FKNN created by the authors of the "A fuzzy KNN-based model for significant wave height prediction in large lakes" paper. (http://bit.ly/2uu1OKp)

Requirements of a Machine Learning model

The requirements of the ML model are usually represented by a specialized technical document, which formalizes the expectations related to the corresponding project, namely the implementation of this ML model in a programming language and its application on actual data. For example, a requirements document may define the particular data that is involved in the usage of the model. For example:

- The ML system shall make use of data contained in the Transactions table of the Cashflow database, as well as data in the Customers-related tables of the Customers database.

- The ML system's dataset shall have data for at least six months of transactions.

- The ML system shall yield both predictions and confidence scores for all new data it is given, once it has been trained.

In addition, it may contain performance-specific information such as:

- The ML system shall demonstrate an accuracy rate of at least 85%.

- The ML system shall have an F1 score of at least 0.9 for the class of interest.

- The ML system shall have performance that doesn't deviate more than 5 percentage points, in a 10-fold cross-validation setting.

- The ML system shall not need more than 0.5 seconds and no more than 1MB of RAM to make a prediction.

The requirements document may go into more depth about the function of the ML system too:

- The ML system shall be able to handle noisy data, such as containing missing values and outliers.

- The ML system shall have proper documentation going over its various functions so that a non-data scientist can understand its high-level functionality and usage.

- The ML system's code shall have sufficient comments in it so that it can be understood by an entry-level data scientist.

Naturally, the more detailed this document, the better the chances of the project being successful, especially if the organization is new at applying data science. Also, it makes the whole experience of working on this project more efficient and less stressful. Note that the requirements document may not always be present since not everyone knows how to write one. In more complex problems, it's best to insist on creating one as it can greatly facilitate the communication between the non-technical members of the team with the technical ones. Usually, a data modeler can assist significantly in this, though the data scientist needs to do his part too.

Implementing a Machine Learning model in Julia using functions

FKNN algorithm key points

Now we are ready to implement the FKNN ML model using Julia. For this use case, we'll utilize the flowchart available in the "A fuzzy KNN-based model for significant wave height prediction in large lakes" paper since it's adequately comprehensive.

First of all, all the membership variables take values between 0 and 1, while their sum is equal to 1, which makes intuitive sense since no single data point can belong to a class more than 100%, while the total "belongingness" cannot exceed 100% either. It follows that the total membership values across all data points, for a given class, has to be between 0 and n, where n is the total number of data points. The membership variable is usually denoted by μ (Greek letter equivalent to m) or u as it's easier to type for people who don't have a Greek keyboard activated in their computer.

Also, the membership of a data point is calculated using the formula:

$$u_i(x) = \frac{\sum_{j=1}^{K}\left(\dfrac{u_{ij}}{||x-x_j||^{\frac{2}{(m-1)}}}\right)}{\sum_{j=1}^{K}\left(\dfrac{1}{||x-x_j||^{\frac{2}{(m-1)}}}\right)}$$

where K is the number of neighbors considered, $||x\text{-}x_j||$ is the Euclidean distance between point x and neighbor x_j, m is a constant that's always greater than 1, and u_{ij} is the membership of data point j in relation to class i.

Note that the whole expression $1/(||x\text{-}x_j||^{2/(m-1)})$ is the similarity of a given data point x to another data point x_j that's part of the training set since it's 1 divided by a form of distance between the two data points. Depicting this similarity as $s_j(x)$, the above equation can be rewritten as:

$u_i(x) = \Sigma(u_{ij} * s_j(x)) / \Sigma(s_j(x))$, where each sum Σ is across all j values (neighbor indexes).

How we define u_{ij} is up to us and there are various ways of doing so. In the article it's not mentioned explicitly. Still, as long as the membership values sum up to 1.0, for every given data point, it is a valid possibility of the membership function. You can view the memberships of a data point for the various classes as proportions, and the latter always add up to one. Here we'll use the most straightforward way of assessing u_{ij} which is the following basic algorithm:

1. For a given data point j in the training set, find all the K neighbors of it (let's call this X_k and y_k) for a given class i
2. Count how many data points of class i exist among the X_k and y_k
3. Divide this number by the total number of data points in the neighborhood (i.e. K)
4. Assign that value to u_{ij}
5. Repeat steps 2-4 for all classes
6. Repeat steps 1-6 for all data points in the training set

FKNN Julia Code

We can code the above algorithm in Julia as follows. First of all, we need to define some auxiliary functions. We can also start from the main method and gradually go into more and more task-specific functions. The similarity function is the most foundational, which is based on a parametric form of the Euclidean distance. Naturally, the original formula could be refined further, something done in its implementation through the *similarity()* function. See the corresponding .jl file for this and every other function developed for this algorithm.

Next, we need to have a way to get the K nearest neighbors of a data point, which is slightly different for the training and the testing phases, since the former case we need just the indexes of data points in each class, while in the latter case we also need the distances of these data points. So, it's best to implement this as two distinct functions: *FindKNearestNeighborsTraining()* and *FindKNearestNeighborsTesting()*.

Afterward, we need to find the classes of these neighbors we have identified, which although straightforward, it's best to have a dedicated function, namely the *FindNeighborsInEachClass()*. The output of this function is twofold: the binary matrix of the indexes, which shows whether a particular neighbor belongs to a particular class or not, and the number of neighbors for each class in total.

Next, we need to figure out the memberships of each data point in the training set, as well as each one in the testing set, with respect to the various classes. This is accomplished through two somewhat similar functions: *MembershipTraining()* and *MembershipTesting()*.

Finally, we have the *DecisionAndConfidence()* function, which decides which class label to use for the classification and how confident the system is for that decision. The latter part is not in the original algorithm, but it's quite easy to implement and quite useful too, as we'll see later on. After all, it's much easier to perform some sensitivity analysis on the algorithm if we have a confidence vector to accompany the predictions one.

Next, we have the main methods of the FKNN implementation. The first one is the training of the algorithm—the calculation of the membership matrix M, the class vector Q, as well as the majority class, which is accomplished by the *fknn_train()* function. The majority class is not essential, but it can be quite useful for certain values of the m parameter that lead to an awkward situation whereby all classes have an estimated membership of 0. When this happens, the majority class is selected since all potential classification options are equally likely.

The other main method is the testing of the algorithm—the application of the membership matrix to new data points, something undertaken by the *fknn_apply()* function. The outputs of this function are the predictions vector Y and the corresponding confidence vector C.

In case you are wondering why model the whole thing using functions, instead of some other ways like programming classes or metaprogramming, let's remember that Julia is a functional language so programming functions is one of its strong suits.

About the FKNN implementation

Naturally, the above is just one possible implementation of the FKNN algorithm. It's quite likely that this is not the same one the authors of the research paper created. However, someone could argue that even a sub-optimal implementation in Julia is bound to have better performance computationally than an optimal implementation in some other high-level language like MATLAB, which is often used for prototyping in research work. Also, the fact that in this particular implementation, there is a confidence score in the output, makes it a more useful method.

In addition, this algorithm, although more advanced than the conventional kNN one, is not always better. In fact, from a computational resources standpoint, it is worse as it requires more computational resources to function, particularly when it comes to the calculation of the membership matrix. Still, it's a useful alternative to consider, especially for a complex data science problem, where this algorithm is bound to perform better than the conventional kNN one, in terms of accuracy rate and F1 score.

Testing a Machine Learning model

Testing a ML model like FKNN is fairly straightforward once all the previous steps are completed. Namely, we'll need to ensure that the FKNN model has a decent score in terms of accuracy rate or some other metric such as an F1 score, in a new dataset that it hasn't been exposed to during training (test set). Also, the performance of the FKNN model needs to be consistent across different test sets. Both of these requirements are important as they guard against high bias and high variance, respectively.

So, after we split the dataset in a number of partitions such as five, we can test the model to see if it performs as well as it should. Then we run a series of five tests using the previous partitions to ensure the performance is on par with our expectations. We can perform all this using the following code:

```
indexes = randperm(N) # indexes of a random permutation of the N indexes
                corresponding to the whole dataset

n = div(N, K) # number of data points in each partition

X_random = XX[indexes,:];

y_random = y[indexes];

P = Array{Any}(undef, K) # initialization of the partitions array

a = 1; # starting point of each partition

b = n; # ending point of each partition

for i = 1:(K-1) # assign n data points for each one of the first K-1 partitions

  P[i] = Array{Any}(undef, 2)

  P[i][1] = X_random[a:b,:]

  P[i][2] = y_random[a:b]

  a += n

  b += n

end

P[K] = Array{Any}(undef, 2) # initialization of the last partition
```

```
P[K][1] = X_random[a:end,:] # assign remaining data points in the last partition
        (feature data)

P[K][2] = y_random[a:end]; # assign remaining data points in the last partition
        (target data)

F1 = Array{Float64}(undef, K) # initialize F1 score array for the K partitions

for i = 1:K # repeat for each partition

    ind = setdiff((1:K), i) # indexes of all remaining partitions (to be used for
            training set)


    X_train = P[ind[1]][1]

    y_train = P[ind[1]][2]


    for j = 2:(K-1)

    X_train = vcat(X_train, P[ind[j]][1])

    y_train = vcat(y_train, P[ind[j]][2])

end


X_test = P[i][1]

y_test = P[i][2]


M, Q, mc = fknn_train(X_train, y_train, k, m) # training the fknn algo

yy = fknn_apply(X_test, X_train, k, M, Q, mc, m)[1] # applying the fknn
        algorithm


F1[i] = F1_score(yy, y_test) # calculating the F1 score for i-th fold of KFCV
end
```

Finally, we can perform ROC analysis to ensure that the system is stable and that the confidence score is meaningful, which can be done using the ROC package as follows:

```
train = indexes[1:N_training]; # first, get the training set indexes (using the
        previously randomized indexes vector)

test = indexes[(N_training+1):end]; # same for the testing set indexes

X_train, y_train = XX[train,:], y[train]; # get the feature and target data for the
        training set

X_test, y_test = XX[test,:], y[test]; # same for the testing set

M, Q, mc = fknn_train(X_train, y_train, k, m) # train the fknn algorithm

yy, c = fknn_apply(X_test, X_train, k, M, Q, mc, m); # apply it; this time get the
        confidence vector as an output

y_with_c = CombinePredictionsWithConfidence(yy, c); # combine y and c in a
        pseudo-probabilistic variable that can be used for ROC analysis

roc_data = roc(y_with_c, y_test, true); # calculate ROC data using ROC
        package

AUC(roc_data) # calculate the area under the ROC curve, a useful evaluation
        metric

plot(roc_data, label = string("FKNN performance for k = ", k, " and m = ", m),
        legend = :bottomright) # plot the ROC curve (Fig. 7.4)
```

Naturally, we can repeat this process for different values of K, to fine-tune the model. If you are feeling up for a challenge, you can explore how different values of *m* affect the outcome.

Showcasing a Machine Learning model

Showcasing a ML model is also relatively simple, though not as easy as it may first seem. It's quite tempting to want to describe every little thing you've done in detail when communicating your work, especially when the audience is knowledgeable and tech-savvy. The key aspects of the model that need to be showcased are:

1. Its functionality, from a birds-eye-view standpoint
2. How well it performs on a dataset, on average
3. How the performance varies for different parameter values, such as for different values of K
4. How stable the performance for different test sets, such as through K-fold cross-validation

For the first aspect, we can make use of the diagram in Figure 7.2. Naturally, if we were to create a diagram from scratch, we'd be more meticulous about it or perhaps hire a professional to help. As for the second and third aspects, we can showcase these on a table or even a basic bar chart. Note that the third aspect is often referred to as "sensitivity analysis," and it's particularly important when analyzing a novel algorithm to ensure that it is reliable enough.

Regarding the fourth aspect, we can use a graphic such as a line chart to show that the model is stable enough to be reliable (Fig. 7.3). A classical case is the ROC curve, which shows how the model performs for various scenarios (Fig. 7.4). ROC analysis is applicable in the case of a binary target variable, however.

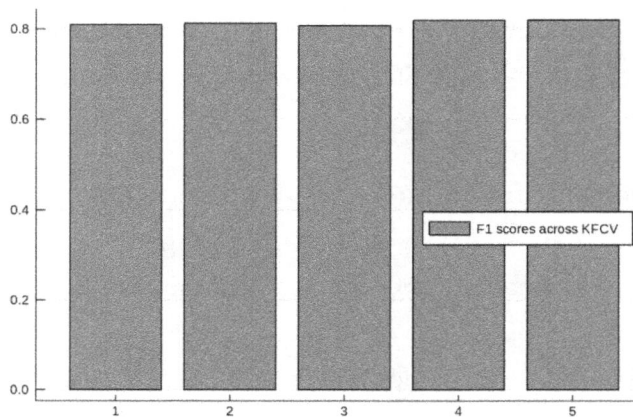

Figure 7.3. Plot denoting the performance of the FKNN algorithm on the Wine Quality dataset for different partitions of the dataset.

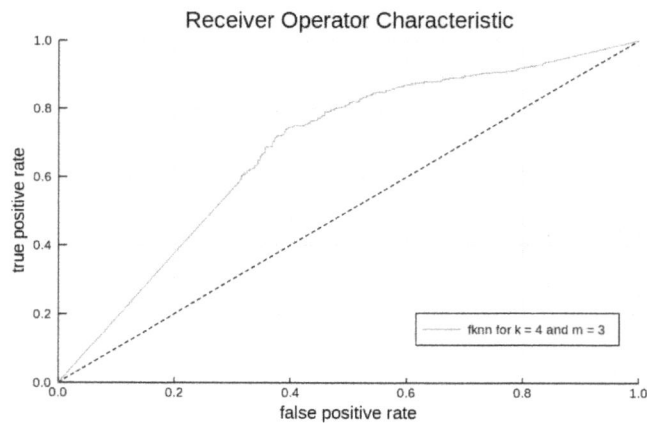

Figure 7.4. ROC plot denoting the performance of the FKNN algorithm on the Wines Quality dataset, for various confidence levels, corresponding to various false-positive rates.

Useful considerations

Based on this example, it may seem straightforward to build a Machine Learning model from scratch, but there are certain things that you need to consider when doing so, to ensure everything goes smoothly. First of all, not all Machine Learning models built are going to derive from some other model. As a result, additional work may be required to build them properly and make sure that they run in a variety of operational settings. So, it's best to build such models when you are not limited by a strict deadline, as rushing the development of such a model will have undesirable results.

What's more, if you decide to build a variant of an existing model, you should consider using preexisting code, such as from a package, and extend it accordingly. This way, not only will you save time, but you will also improve your coding since professional coders create some of these packages. However, building from scratch allowed us to show how easy it is to prototype in Julia when it comes to Machine Learning systems.

In addition, when developing a model based on a new Machine Learning algorithm (so you have to build everything from scratch), it's best to implement it piece by piece. This way, you can ensure that every component works as expected, thereby minimizing any potential issues when you test the model as a whole. Also, you can come up with a series of potential tests to run on these components (called unit testing). As a bonus, the various parts of the Machine Learning model could be used in other projects, saving you time and effort.

Moreover, it's fundamental to make use of comments when developing a custom model in Machine Learning. What you do in the script may not be so obvious, and you may have a hard time figuring out what you've done if you look at it a few weeks later. Also, if you plan to share this code with others, it's good to make sure that the code is clean and easy to read. All this is bound to make the model more future-proof and easier to upgrade when necessary. In addition, it makes sense to document more complex scripts, including their usage.

Furthermore, when developing a model from scratch, especially if it's a complex model, it's best to map out its various stages, as well as the inputs and outputs, which facilitates testing while avoiding lots of potential errors. Also, it would be more feasible this way to optimize the various components and ensure that the system has good performance overall. It's also easier to maintain if there is a schematic describing its various components.

Finally, when developing a custom-made model in Julia, it's important to be aware of the changes in the language as new releases become available. Functions tend to change as the language evolves, so the script you've written for the previous model needs to be maintained. The same goes for any dependencies since some packages take a while to adapt to the new language release. In other words, if you want the model you've created to be useful in the foreseeable future, it needs to be maintained.

Summary

- The design of the model is essential for understanding how the model works before we delve into the details. It's also useful for showcasing the model afterward.

- The requirements document is also quite useful, though it's not always available. It demonstrates what the model should and shouldn't do clearly and comprehensively so that even not-technical stakeholders of the project can understand.

- Implementing the ML model involves "translating" the pseudo-code into actual programming code. When doing so in Julia, it is best to make use of a variety of functions to mitigate the chances of errors and the time needed to fix them. Also, code that involves lots of functions is easier to maintain and possibly reuse in other ML projects.

- Testing the ML model involves establishing its performance level, in terms of a metric that we need to introduce beforehand, such as an F1 score, its sensitivity to the model's high-level parameters (aka hyper-parameters, such as K and m), and its reliability in terms of how consistent the performance is across different test sets. The latter is a set that has the same features as the training set, but the model hasn't accessed data from it during the training phase.

- ROC analysis is particularly useful when the confidence scores for the various predictions are also available, as in the case of this modified version of FKNN. ROC analysis can help us delve deeper into the sensitivity analysis of a model, when a binary classification task is involved.

- The training data is only used in the training phase for the ML model. However, this doesn't apply to the FKNN model due to the nature of the algorithm. Even though during the training phase, the training data is used for calculating the membership matrix, the features of the training dataset are also used in the testing phase to calculate the distances involved.

- Showcasing the ML model is very important as it involves demonstrating that the model was worth your time and that it can be used moving forward.

- When building a ML model from scratch, remember these tips:

  o Not all models derive from preexisting ones.
  o It's useful to use code from programming libraries when possible.

o It's best to implement new models piece by piece, using comments is fundamental, especially when performing complex processes.

o Mapping out the whole development procedure can be very helpful.

o Being aware of the language's changes over different versions is something to be mindful of when building a model from scratch in Julia.

Questions

1. How would you parametrize the model showcased in this chapter even further?

2. What do you need to do regularly in order to maintain the usefulness of a custom-made model like this one?

3. Could you use pre-made code (e.g. from a different codebase) in your custom-made model, to save time? What considerations do you need to have when doing so?

4. How would you prove to someone that this custom-made model is any good, for the task it is designed for?

5. Can a custom-made model be comprised of other models, such as in an ensemble setting?

6. When would you use a programming library, and what considerations would you have when doing so?

7. Based on what has been discussed in this chapter, would a report resembling the paper used here be effective? How would you supplement it to make it more comprehensive and more comprehensible?

8. When would you avoid relying on an academic paper showcasing a novel algorithm, relevant to Machine Learning? Why?

9. Could you use ROC analysis in the case of a multi-class classification problem?

CHAPTER 8

Machine Learning Methods for Dimensionality Reduction

Dimensionality reduction is an essential part of data engineering, particularly feature engineering. Although there are many robust statistical methods such as Principal Components Analysis (PCA) Independent Components Analysis (ICA), and Factor Analysis (FA), there are also a few within the Machine Learning ecosystem. Dimensionality reduction is a type of unsupervised learning, as only the feature data is used in most cases, even if a target variable is available.

Dimensionality reduction plays an important role in many data science projects, particularly when visualization is required. As a result, it's important to know about it, even if you are dealing with datasets comprising a relatively small number of variables. After all, the advantages of dimensionality reduction spread over many use cases, not just datasets of high dimensionality.

In this chapter, we'll explore various aspects of Machine Learning based dimensionality reduction. We'll start with the curse of dimensionality and the value of dimensionality reduction. Then we'll go on to explore a taxonomy of dimensionality reduction methods based on the nature of the underlying optimization algorithms. Afterward, we'll look at three powerful and fairly popular dimensionality reduction processes that you can utilize in Julia: Locally Linear Embedding (LLE), Isomap, and UMAP.

"Curse of Dimensionality"

As the dimensionality of a dataset increases, we notice something peculiar. Namely, the distances among the data points tend to become more or less the same—their variance doesn't grow, which makes the distance metric useless, obscuring the structure of the data. As a result, models making use of distances or any similarity metric based on distances tend to fail. This is quite common with clustering methods, but certain predictive analytics models exhibit the same issue, too, such as kNN and other transductive algorithms.

The curse of dimensionality is quite common, particularly in cases of text analytics, image analysis, and sensor data, where the number of variables involved is inevitably high. Also, combining different datasets leads to high dimensionality feature spaces, resulting in the curse of dimensionality. Moreover, when dealing with lots of categorical variables, we end up breaking them up into a series of binary (dummy) variables to use them in our models. This binarization process often makes the dimensionality of the dataset increase dramatically.

It is not clear at what level of dimensionality this issue appears. However, we can say that the higher the dimensionality, the more likely we are to have issues. Also, if the number of data points is relatively small, the curse of dimensionality issue worsens significantly. As a rule of thumb, you need to have at least twice as many data points as there are features. Fortunately, there is an abundance of data points in most modern datasets. Yet, this abundance makes the features abundant too, which is what triggers the curse of dimensionality issue again.

Value of dimensionality reduction

The value of dimensionality reduction is unmistakably immense, especially today when large datasets are commonplace—often, we need to reduce dimensionality before being able to build a robust predictive model. That is, dimensionality reduction is an antidote to the curse of dimensionality.

In addition, sometimes the original data has *collinearity* issues, making certain data models not perform well, translating into redundancy and higher computational costs. So, removing or combining those features not only makes sense, but it's also necessary in many cases. After all, certain models work better if the features are linearly independent of each other, which is ensured by many dimensionality reduction methods.

What's more, dimensionality reduction aids data visualization. Most datasets cannot be visualized without first reducing their dimensionality. Even a four-dimensional dataset would be difficult to visualize, requiring several plots, usually of different pairs of variables. As visuals are an important part of data science work, particularly EDA, dimensionality reduction is a powerful tool that's indispensable for any data science project. Also, certain methods like clustering are too abstract to communicate, making visuals an absolute necessity.

Finally, by reducing the dimensionality of a dataset, we can conserve computational resources, requiring less storage space and RAM, as well as less CPU and GPU power. As a result, this means that computations on the reduced feature set are going to be faster for a given machine, saving us time in our data science projects. Also, in the case of cloud computing, this can translate into lowering the cost of the whole project since cloud usage costs can add up when dealing with large datasets.

Main methods for dimensionality reduction

There are many ways to perform dimensionality reduction. These can be organized into two mutually exclusive groups, depending on the underlying optimization algorithms involved: deterministic and stochastic. Dimensionality reduction involves finding the best possible representation of the original features in smaller dimensionality, given certain restrictions. This new representation is often referred to as *embeddings*, or more accurately, meta-features, since it usually involves some kind of combination of the original features.

Deterministic

These methods are based on matrix factorization and rely on linear algebra techniques to reduce dimensionality. Deterministic methods include algorithms like PCA, Independent Components Analysis, Linear Discriminant Analysis (LDA), Factor Analysis, Non-Negative Matrix Factorization (NNMF), and certain feature selection techniques. The common characteristic of all these methods is that they always yield the same output for the same inputs, making the end result predetermined (hence the name "deterministic").

Most of the deterministic dimensionality reduction methods are statistical in nature and are ideal for smaller datasets, both in terms of dimensions and in number of data points. Since many of these methods involve complex matrix operations, such as finding the eigenvalues or the inverse of a matrix, they don't scale very well. However, they are handy when dealing with samples of a dataset or not too large dimensionality. As these methods are not directly related to Machine Learning, we won't examine them in this chapter. However, you are encouraged to explore them on your own as there are useful packages in Julia covering most of them.

Stochastic

These methods are based on neighbor graphs and rely on optimization methods, particularly stochastic ones, to reduce dimensionality. This set of methods includes algorithms such as Isomap, Uniform Manifold Approximation and Projection (UMAP), Laplacian Eigenmaps, Locally Linear Embedding (LLE), and T-SNE.

Most of these methods are based on manifolds for figuring out the structure of the dataset, while others, such as T-SNE, rely on other heuristics. The vast majority of the stochastic methods are under the Machine Learning umbrella and are the ones we'll focus on in the next section. Due to the nature of the optimization algorithms utilized in these methods, the results are somewhat different every time they run. Hence the term "stochastic" as a way to describe them. Also, these methods fall generally into one of two categories: those focusing on the local geometry of the data and those geared toward the global geometry.

Local (geometry) methods include LLE, T-SNE, and Laplacian Eigenmaps. Although useful in some areas such as data visualization, they are not ideal as they lose a lot of information related to the global geometry of the dataset. This issue is tackled in the global (geometry) methods, which include Isomap, Multidimensional Scaling (MDS), and UMAP. This distinction is non-existent in the deterministic methods as methods of that category are all global. You can view this whole taxonomy of dimensionality reduction methods in Fig. 8.1.

Dimensionality Reduction methods

Deterministic   Stochastic

Global   Local

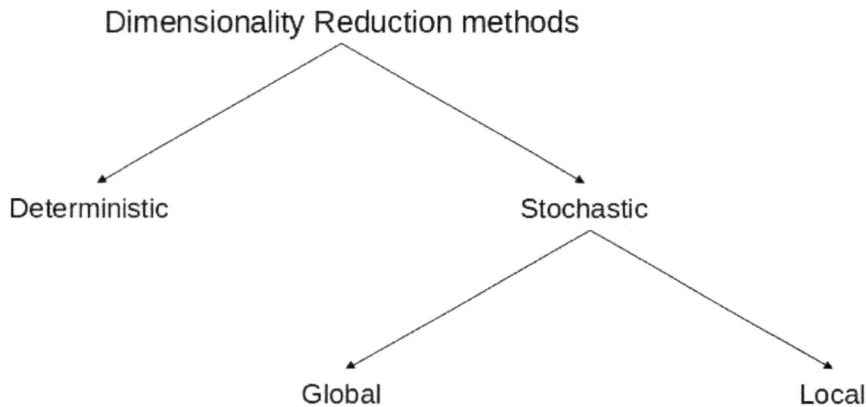

Figure 8.1. Dimensionality reduction methods

Machine Learning methods for dimensionality reduction

Locally Linear Embedding (LLE)

Locally Linear Embedding is an interesting method that attempts to create meta-features using local structure information, based on the k nearest neighbors of each data point. Its niche is that it exploits the local symmetries of linear reconstructions, which is why the manifolds it generates are not a good representation of the dataset as a whole, even if they capture the local structure. You can learn more about the details of the LLE method in this paper, downloaded in PDF format: http://bit.ly/392s4v6.

To apply LLE in Julia, you need to use the *fit()* function of the MainfoldLearning package as follows:

```
fit(LLE, XX; k = nn, maxoutdim = nd)
```

where *XX* is the feature set to be reduced, k is the number of neighbors parameter (default is 12), and *maxoutdim* is the (maximum) number dimensions in the output (such as 2). Note that if the number of dimensions in the output (nd) is smaller than the number of neighbors (k), as in this case, you will get a warning saying that the method will employ regularization. That's nothing to worry about as it's not an error in any way.

The resulting data structure is an LLE model that fits this data. To access the meta-features themselves you need to apply the *transform()* function:

    transform(DRM)

where *DRM* is the dimensionality reduction model we've built previously.

When applying LLE for the wine quality dataset, reducing the dimensionality to two, we obtain the plot depicted in Fig. 8.2. The meta-features the method created are of very similar scale, although the dataset looks peculiar due to local dimensionality reduction methods like LLE. Also, you can tell from just a glance that the meta-features the method yields are quite independent of each other.

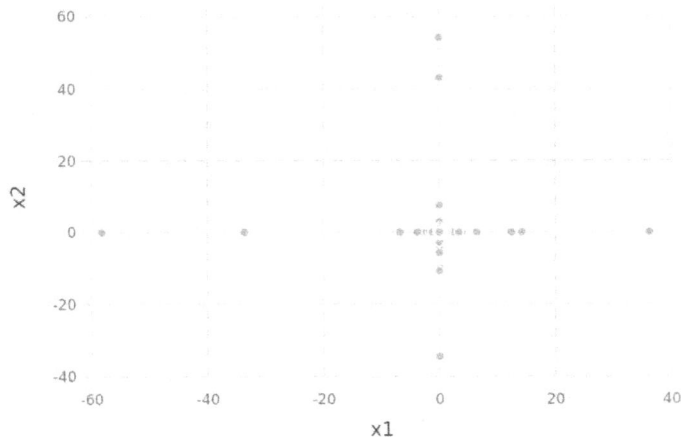

Figure 8.2. Wine quality dataset reduced to two dimensions using the LLE method.

Isomap

Isomap makes use of manifolds to create low-dimensionality embeddings using the k nearest neighbors of each data point, all while preserving the global structure of the dataset. It computes non-linear meta-features that have a similar structure to the original dataset. Tenenbaum and his team in 2000 created Isomap as part of Stanford's research. The key advantage of this method is its simplicity, as reflected in its single parameter, namely the number of neighbors, k. You can apply Isomap in Julia using the *fit()* function of the MainfoldLearning package like previously:

fit(Isomap, XX; k = nn, maxoutdim = nd)

where *XX* is the feature set to be reduced, *k* is the number of neighbors parameter (default is 12), and *maxoutdim* is the (maximum) number dimensions in the output (such as 2).

Note that the *transform()* function mentioned in the documentation of this package as a way to process the data directly is deprecated, so it's best to use *fit()* instead. Also, you can use the same function with other manifold-related methods instead of Isomap, such as Laplacian maps. You can read more about the various model supported by the ManifoldLearning package in its documentation page: https://manifoldlearningjl.readthedocs.io.

The resulting model that the *fit()* function yields is a data structure comprising the model parameters, the components (data points) used, the eigenvalues of the reduced feature set, and the meta-features (labeled as "projection"). Note that both the original features (XX matrix) and the meta-features are represented as rows instead of columns. As a result, it's best to apply the transpose operator (') both before and after using Isomap.

When applying Isomap for the wine quality dataset, reducing the dimensionality to two, we obtain the plot depicted in Fig. 8.3. Interestingly, the meta-features the method created are of the same scale. Also, the reduced dataset seems to be fairly compact (no outliers at all).

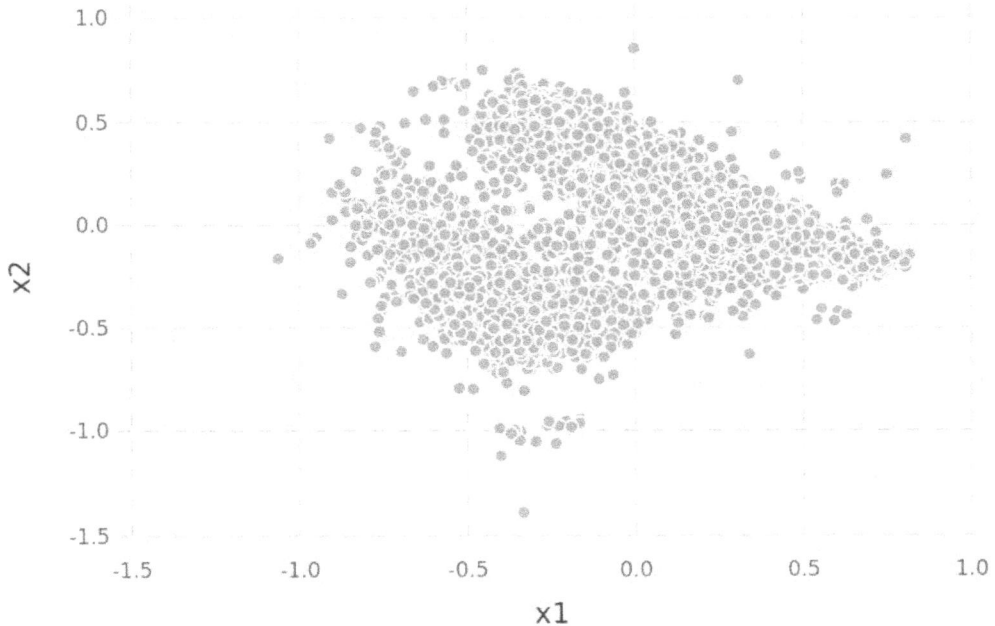

Figure 8.3. Wine quality dataset reduced to two dimensions using the Isomap method.

UMAP

Just like Isomap and LLE, UMAP has strong mathematical foundations, namely topology, and is excellent in terms of efficiency and usability. If you know UMAP well enough, using any other stochastic dimensionality reduction method seems like a terrible waste of time, especially when using larger datasets where other methods take a very long time. However, for optimal results, UMAP is best combined with some other dimensionality reduction methods, such as PCA, so that it doesn't take that long to function. Ideally, you'd apply

UMAP on the meta-features PCA creates instead of the original data, getting the best of both worlds. You can use this method by applying the *umap()* function of the UMAP package in the following way:

umap(XX, nd; n_neighbors=nn, metric=metric, min_dist=md)

where *XX* is the feature set to be reduced, *nd* is the number of dimensions (meta-features) in the reduced feature set, *n_neighbors* is the number of neighbors parameter, *metric* is the distance metric to be used, and *min_dist* is the minimum distance parameter.

Just like with the LLE and Isomap methods, both the original features (XX matrix) and the meta-features are represented as rows instead of columns so that the transpose operator would be useful here too. You can view the application of UMAP on the wine quality dataset in Fig. 8.4. Interestingly the meta-features created are not scaled as there are a few data points that the method yields as outliers.

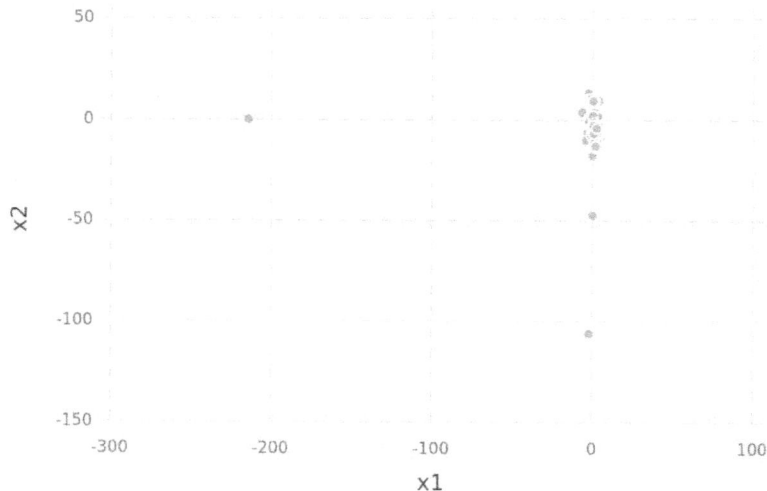

Figure 8.4. Wine quality dataset reduced to two dimensions using the UMAP method.

You can learn more about UMAP in the package's documentation page at https://bit.ly/34L7gH7, as well as in this paper: https://bit.ly/2Kh1hjK.

Other

Beyond these three methods, there are other dimensionality reduction techniques under the Machine Learning umbrella. For example, Self-organizing maps (SOMs) and Autoencoders both perform the same task, using a graph-based approach that doesn't involve manifolds like the aforementioned methods. In the case of SOMs, the network itself is the reduced dataset, while in the case of autoencoders, an ANN is used with its innermost layer being the reduced feature set. However, due to the complexity of these methods and the fact that they require more advanced knowledge, we won't look at them in detail.

Another method that's worth mentioning is T-SNE. Although this method was revolutionary when it came out, its popularity waned over the years when newer and better methods like UMAP came about. Also, since the corresponding package in Julia is overly complex and contains poor documentation, T-SNE is a substandard option for dimensionality reduction. Besides, T-SNE's main use case is reducing the dimensionality of the data for better visualization. Other dimensionality reduction methods, however, can provide use cases that go beyond data visualization, making them more versatile options.

Useful considerations

Although dimensionality reduction seems straightforward, there is a bit more to it than that. After all, just like classification, regression, and other data science methodologies, there are lots of subtleties to it that you need to keep in mind.

First of all, dimensionality reduction almost always involves some loss of information. The new features may capture most of the information from the original dataset, but some of it inevitably slips away. Although this may not be an issue for ML applications, it's something worth remembering. The value of a reduced feature set has to come at a cost. This cost, however, is usually worth paying since the benefits dimensionality reduction provides outweigh the cost of information loss.

Also, most ML methods for dimensionality reduction don't help you decide how many meta-features to use, making the selection of the corresponding parameter a challenge. Additionally, they are opaque in another respect, namely in linking the meta-features with the original features, something that is possible with methods like PCA. As a result, we cannot know how much each one of the original features contributes to the meta-features these methods yield.

What's more, as with most dimensionality reduction methods, the ones here require normalization before applying. Unlike PCA, however, these methods don't require standardization, since any normalization method would be fine. After all, these methods rely on distances, so they need everything to be comparable in range to work well.

In addition, the dimensionality reduction explored here is under the umbrella of unsupervised learning as it doesn't involve the target variable. However, there are methods for reducing dimensionality, such as feature selection, that use the target variable, which is why some people consider these methods supervised learning.

Moreover, the dimensionality reduction methods we viewed are best applied on the whole dataset, both training and testing, before using the reduced feature set for a predictive model. This is a subtle point, but it's vital as it can skew the performance of your model considerably. Naturally, you'd need to apply the same normalization model on the whole dataset too to ensure that both partitions of the dataset are of the same scale. If, however, you feel that you shouldn't use the testing set at all, you can rely on the training set only to create the dimensionality reduction model and then apply the latter on the test set.

Furthermore, dimensionality reduction is often necessary when performing clustering, since many clustering algorithms don't perform well in a high dimensionality space. However, you don't need to reduce the feature set to two dimensions, unless it is just for visualization. The methods discussed work well with other dimensionality options too (e.g. 5 dimensions).

Finally, the meta-features dimensionality reduction methods yield are often in need of normalization or even further processing. In the case of UMAP, for example, we saw that some data points took extreme values, so they would need to be treated accordingly before the effective use of the reduced feature set.

Summary

- The curse of dimensionality involves the distance metrics ceasing to yield useful information when the dimensionality of a dataset increases.

- Dimensionality reduction is very important, especially today, as a way to tackle the curse of dimensionality. Dimensionality reduction can help ensure a model is trained properly, while it can also aid EDA, help conserve computational resources, and save time.

- The various dimensionality reduction methods can be categorized as deterministic and stochastic, depending on the nature of the underlying optimization algorithms. Most deterministic methods tend to be statistical in nature, while the Machine Learning related methods tend to be stochastic.

- Modern dimensionality reduction methods under the Machine Learning umbrella are related to manifolds, such as LLE, Isomap, UMAP, and Laplacian Eigenmaps.

- Certain Machine Learning methods for dimensionality reduction focus on the local geometry of the dataset (e.g. LLE), while others take into account global geometry (e.g. UMAP).

- Feature selection methods are also quite useful as they are easy to use, interpret, and tend to be faster than all of the other methods discussed. Most feature selection methods rely on some kind of heuristic for evaluating the features at hand.

- The key packages in Julia for Machine Learning-based dimensionality reduction are ManifoldLearning and UMAP. Both have good documentation and are easy to use, while the ManifoldLearning includes a variety of methods, such as LLE, Isomap, and Laplacian Eigenmap.

- Dimensionality reduction always involves some loss of information, something that's often necessary as it helps handle the data better. Also, it's best to scale the features before applying dimensionality reduction.

- Although most dimensionality reduction methods involve just the features of the dataset, some of them involve the target variable too. In cases like this, dimensionality reduction is labeled as a supervised learning approach, even if it is under the unsupervised learning umbrella.

- Oftentimes, the meta-features deriving from a dimensionality reduction method require normalization or even further processing.

Questions

1. What's the advantage of UMAP, LLE and Isomap over PCA?

2. Would you ever use a dimensionality reduction method on a dataset with fairly few variables? Explain.

3. What's the main disadvantage of UMAP, LLE and Isomap in relation to PCA and ICA?

4. Are there any AI-based dimensionality reduction methods? If so, which ones?

5. How do UMAP, LLE and Isomap compare to feature selection methods based on the target variable?

6. What's UMAP's main advantage over Isomap?

7. Could different dimensionality reduction methods be used in tandem? What would be a good strategy for that?

CHAPTER 9

Machine Learning and Julia Additional Topics

Although the Machine Learning programs we discussed in the previous chapters are fairly straightforward and their implementation in Julia intuitive enough, there are other topics essential for Machine Learning practitioners, independent of the programming language. Also, there are certain aspects of Machine Learning that gain a different meaning with a Julia perspective.

In this chapter, we'll look at these topics, starting with over-fitting, an issue that tends to break a Machine Learning model if it's not attended. The bias-variance trade-off is something we'll explore next since it's closely linked to over-fitting, and it's only through comprehending it that you can develop an intuition on what makes a model good. Afterward, we'll look into differentiating between Machine Learning and Statistics as well as using models of the latter framework in parallel with Machine Learning models.

We'll continue by exploring the value of heuristics, a topic that doesn't receive enough attention, even though it is paramount in any novel Machine Learning process. We'll then cover Machine Learning ensembles and how they add value to data science projects, as well as how a Machine Learning model can be optimized.

Over-fitting

Over-fitting is an issue that plagues even the most promising models, especially those under the Machine Learning umbrella. As the name suggests, it has to do with fitting the model too much—so much so that it fails to perform well when applied to new data. A good model is not one that understands the training data perfectly, but rather one that can work well with any new data.

Getting the model to perform well with new data isn't always easy—all we have is training data and sometimes corresponding labels. As a result, a model has to actually learn the underlying patterns in existing data to perform well with new data, including how this data relates to what it's trying to predict, in the case of predictive models. This is a process known as *generalization* since it involves generalizing based on the patterns it observes and making inferences based on this generalization.

If the model tries too hard to find all possible relationships on the training data, it fails to generalize properly since it thinks that the training data is all there is to the whole problem. In other words, it neglects to address possible variations from that data, something not uncommon in new datasets. That's when over-fitting creeps in, and unless the model is retrained (this time with less zest!), it is bound to make lousy predictions.

In practice, over-fitting is observed when the performance of the model fluctuates significantly. As a rule of thumb, if the performance metric, such as accuracy rate, F1 score, or MSE, has a range of more than 25% of its average value, it's a sign that it's over-fit. More often than not, this performance is measured on different partitions of the training data, as in the case of K-fold cross-validation with a K value of 5 or so. However, if you perform random sampling yourself several times, you can obtain something equally useful, though you may need to take more samples to be sure.

Graphically, over-fitting looks like a line that follows the training data very closely, trying to go through every data point. You may recall from statistical models that a straight line is often utilized, such as in the case of a regression problem. However, any kind of line can be used, and as Machine Learning is quite liberal in the ways it approximates the data, it is possible to employ an overly zealous approximation. When plotted, this looks like a very curvy line that passes through most, if not all, of the data points on the training set. Naturally, since the test set is bound to be at least somewhat different than the training set, this line is not going to pass through the points of the testing set too, resulting in errors and always unstable performance, something you'd want to avoid in all your models.

Although an over-fit model may perform well in some cases, such as when the test set is very similar to the training set, it is not reliable. A lighter model with humbler performance is preferred as it's bound to yield more robust results, as we'll see in the next section.

Finally, over-fitting is something that usually plagues more sophisticated models. That's why it is often the case that we guard against it in ANNs and other complex models. A simpler model is prone to over-fit too, but it's less likely and more avoidable. Avoiding over-fitting is one of the reasons why we often start a data science project with a simple model and gradually move to more complex ones.

Bias-variance trade-off

The bias-variance trade-off is by far, one of the most important theoretical aspects of predictive analytics, and its applications are found everywhere in data science, including Machine Learning. However, the bias-variance trade-off applies to predictive analytics models, and is essential when optimizing them.

In a nutshell, the bias-variance trade-off states that the less biased a model, the more variance it will have and vice-versa. Bias is the characteristic of a model that has to do with its general competence—how well it approximates the target variable on average. As for variance, it has to do with the particular competence of the model—how well it does when it comes to specifics. In general, you'd want both of these to be low, since low bias means that its predictions are not far from where they're supposed to be, on average, while low variance means that it doesn't vary too much from the specific values it is supposed to predict.

So the trade-off between these two is a heuristic rule stating that you can't have both at a great value. If you want to have a model that is completely unbiased, it will have to have high variance and be somewhat unstable, and if you want to have a model that's as reliable as possible with performance that doesn't vary, it is bound to have a high bias. If a model is better than another one both in bias and variance, it is a better model overall.

Variance of a model is different than variance of a variable. The latter is a useful metric of a variable's values, while the former has to do with a model's performance. However, on a deeper level, the two are related, hence the use of the same name for both metrics.

The bias-variance trade-off is essential in understanding models in general, too. This is because it is a statement that helps us connect these two seemingly independent metrics and explore the space of possible models more methodically. Fortunately, the number of models known is fairly limited, so this search is not too time-consuming. In classification problems, this takes the form of comparing performance ROC curves and calculating the areas under them

(AUC metric), by altering one or more of the model's parameters, and changing its classification behavior.

Note that the bias-variance trade-off applies to a particular model at a given time. In other words, every model has its own bias-variance trade-off, which needs to be established before we can optimize the model. This is not too difficult, but it means that we need to tweak its parameters a bit. The more variants of a model we explore (through tweaking its parameters), the more accurately we evaluate the bias-variance trade-off. In classification, this means that the ROC curve is going to be smoother, and the corresponding AUC metric more accurate. The downside is that it takes longer to find all these points, especially when dealing with a larger dataset.

In practice, the bias-variance trade-off appears in the performance results of a model as it's being optimized. After all, when we tweak the model's parameters, we do so to improve its performance. Even though it's clear what constitutes an ideal performance, there isn't usually a consensus as to how much bias and how much variance someone is willing to accept in a predictive model. In classification, the exact amounts are calculated by deploying a cost function, which in essence quantifies the cost of each kind of error. The latter is something more tangible and generally a better proxy of the bias and variance of the model.

It's impossible to overstate the significance of this law and its value in the applications of the analyzed models. It's generally better to understand this better instead of learning a lot of models without comprehending the framework this law provides. Also, if you ever want to develop your own models or combinations of models, the bias-variance trade-off is something you'll need to use to evaluate them properly and utilize them reliably.

Machine Learning versus Statistics

It is important to start drawing a line between the Machine Learning group of models and the statistical group. As mentioned in previous chapters, Machine Learning is a purely data-driven approach to analytics, while Statistics tends to use a mathematical model related to some distributions to make its inferences. Both approaches are valid, but the plethora of assumptions made in a statistical predictive model, for example, makes it less practical. For example, for a linear regression system (typically a statistical model), there are six assumptions about the features. On the contrary, a Machine Learning regressor, such as an SVM or kNN, doesn't need to make any assumptions (even the "features should be of the same scale" is in fact more of a suggestion than anything else). Also, the assumptions made by the statistical models tend to be not just about the target variable but each one of the features involved. If there are plenty of features, it may be time-consuming if not utterly impractical to check all of them, to ensure the assumptions hold true.

Machine Learning also enables you to imagine more about what happens in the data space and possibly come up with your own variations of the existing models or come up with entirely new ones. The fact that now many such models are advanced enough and backed up by a solid mathematical framework makes them even more reliable. Statistics, on the other hand, although mathematically robust, seems to use knowledge from a previous century, while the level of math used is unnecessarily complex and single-sided. In Machine Learning you may encounter lots of geometry, for example, while in Stats all you see are calculations based on ad-hoc distributions, as well as probability computations.

The flashy aspect of Statistics has always been the intuitive plots that accompany the models and the detailed tables containing all the metadata associated with them. These were challenging to compute back in the day, and the information contained in these artifacts made these models valuable and transparent. However, all this fancy stuff can now be done easily with a statistical package in any high-level language, including Julia. Although all the Machine Learning stuff can be done equally easily, there are more options available when it comes to models, plus more expertise is required from the user.

What's more, whatever you learn about Machine Learning you can also apply to Statistics, when it comes to predictive analytics. So, the two fields are not as mutually exclusive as they first seem. In other words, they are connected under the data science umbrella and can benefit each other immensely, as we'll see shortly.

Statistical models in tandem with Machine Learning

After differentiating between these groups of models, we can now use them more deliberately and avoid the confusion that stems from considering them the same thing. So, for data analytics tasks that require high interpretability, statistical models can be used. Also, for tasks that require high performance, Machine Learning models can be the go-to option. What about tasks that require a bit of both, however? In this case, a combination of Statistics and Machine Learning models can be utilized. However, the use of the more interpretable Machine Learning models is also a possibility, such as random forests and boosted trees.

What's more, due to their nature, statistical models tend to make different errors in their predictions than most Machine Learning ones, which is a good thing when combining models since the resulting meta-model has better generalization when created properly. We'll talk about combining different models to form such meta-models in a later section of this chapter.

In addition, you can use a statistical model for feature selection and then apply a Machine Learning model on the resulting feature set. After all, a more compact feature set, comprised of relevant features for the predictive task at hand, is ideal for any predictive model. Certain statistical models excel in that, such as lasso and ridge regression systems, making it possible to perform a powerful feature selection before a more in-depth analysis. If full transparency is not required in the end-result, a Machine Learning model can take over after the selection is made by the statistical model.

Finally, you can use specialized Machine Learning methods to assess whether a dataset is better off being analyzed with a simpler statistical model and when it requires a more sophisticated one, such as an AI-based model. This way, you can get the best of both worlds. But how can you assess the complexity of a dataset so that you know when to use what? Well, that's where heuristics can come in quite handy.

Heuristics

Let's now look at heuristics and how they add value to Machine Learning, as well as where Julia fits. First of all, heuristics in data science are methods and metrics that are entirely data-driven and focus on performing a very specific task in an efficient and scalable manner. They may not always be linked to predictive analytics, though they often tend to help with that in some way. Although heuristics add a lot of value to a data science project, they are unsung heroes since they work behind the scenes, be it in data engineering, inside a predictive model, or even in the way a model's performance is evaluated.

Also, Julia is equipped for all kinds of heuristics. This is fairly evident considering it's a high-level language, so prototyping in it is straightforward. However, the fact that it's also quite fast allows you to test any newly coded heuristic on larger datasets and gauge its performance better and even start using it in your work straightaway. Also, since heuristics are custom-made solutions that usually don't rely on any other programming scripts, coding them in Julia doesn't have any drawbacks like those of coding sophisticated models that make use of libraries that may not yet exist, at least not in a reliable state. You can view three Machine Learning related heuristics developed in Julia in Appendix C.

Moreover, heuristics are easy to prototype if your vision of the task is clear. As they are oriented toward a very specific task, it's not too difficult to develop and implement them. Their minimalist nature makes them ideal for auxiliary tools in all kinds of data science processes and the programming aspect of them is fairly straightforward. That's not to say that developing a new heuristic is easy. It takes a solid understanding of the problem at hand and of algorithm design. Once the algorithm is done, however, developing it into a script, usually in the form of a function, is not too challenging, especially with a language like Julia. Sometimes, however, you may need to make use of other functions, such as the *norm* one from Linear Algebra, if you don't want to code everything from scratch.

Furthermore, heuristics are very modular, making them a versatile tool for many projects, even the ones not related to what they were designed for originally. This is very much aligned with the functional programming paradigm, whereby most programs take the form of functions, which are independent and modular by design. So, heuristics and functional programming languages, like Julia, are quite congruent. This makes heuristics a powerful option, one that can be used in various ways in data analytics projects.

Naturally, as they are also data driven, their compatibility and usefulness in a Machine Learning setting is indisputable.

Finally, heuristics are insightful and can facilitate a data science project in a variety of ways. Practically, this means that if you want to get a feel for a variable or the relationship between two datasets, it's much easier to do using a heuristic rather than building a model for the task. Apart from the training cost, the latter may require the data to be engineered in a specific manner, while you may need to try out several models to obtain the required level of performance. A specialized heuristic could do the same thing on its own, without any additional preparation work. As a bonus, it can do its task faster as it's generally simpler in its functionality.

Machine Learning model ensembles

What about ensembles of Machine Learning models? Ensembles of Machine Learning models are often the best alternative when high performance is required and transparency is not important. The reason is that they are so widespread these days that many data scientists use them without even realizing it—in fact, you have already used them in the previous chapters.

However, with few exceptions, all of the Machine Learning ensembles are black boxes. So, as seen previously, we need to sacrifice transparency for high performance. This is usually inevitable since combining different models often entails the use of a meta-model that uses the outputs of several models as its inputs. As for the aforementioned exceptions, these are the homogeneous tree-based ensembles, such as random forests and boosted trees. Such ensembles offer a decent amount of transparency and better performance, though they may be lacking in the latter compared with other ensembles.

What's more, ensembles take a long time to ready for production, mainly due to the fact that they involve a series of training sessions, one for each of their components. Now, many ensembles that are used today tend to use a simple model under the hood, with the decision tree being the most common. Such models are fast and easy to configure, so their training doesn't take that long. However, these ensembles tend to use a large number of these models, so the training time adds up. Also, when fine-tuning an ensemble, you have a number of parameters to optimize, something that makes the whole process quite time-consuming. We'll talk more about fine-tuning in the next section.

In addition, ensembles can over-fit, particularly if you use a lot of models in them or if these models are over-fit themselves. So, if you build a random forest ensemble and choose the tree depth to be high, such as 15, while the data is fairly simple, such as having only five variables, you can be certain that the decision trees in that ensemble will over-fit, making the ensemble over-fit too. However, the alternative of having an under-fit model, equivalent to high bias, is equally undesirable. So, the bias-variance trade-off doesn't disappear in ensemble models.

Moreover, ensembles can also be heterogeneous, meaning that they comprise of different kinds of models. Such ensembles tend to have better performance overall for the number of models they use. However, you need to be comfortable with each one of these models since they require a different set of parameters and optimizing them may not be a straightforward task. Due to their different nature, these models tend to make different kinds of errors—not overlapping errors. As a result, if you combine these models properly you can end up with a supermodel that gets many more predictions right, since at least some of its components will get each prediction correct.

Finally, most ensembles often make use of some heuristic to determine their final prediction, based on the predictions of their component models. The most common such heuristic is the majority vote, which is probably the simplest heuristic in existence! The majority vote states that whichever prediction has the most models behind it is the one you should go with in the ensemble result. Other ensemble heuristics may include more complex metrics, such as the reliability of the component models. The latter can be used as a weight to make sure that the most reliable models are factored in more prominently. Whatever the case, unless you have a bunch of component models that are all very similar, for optimal performance in the ensemble it makes sense to use some clever heuristic when combining their results.

Optimizing a Machine Learning model

Optimizing a model is not a simple task, particularly when it comes to ensemble models. It entails a lot of fine-tuning parameters to avoid over-fitting as well as under-fitting—a necessary process before the model can be used in production and trusted with new data. On top of all that, it's a process that needs to be undertaken for every unique problem tackled, since different problems are bound to require a different set of parameters for their optimum performance.

Optimizing a model is all about making sure that it behaves consistently across different data samples. It's not enough for it to have good performance on the data it is trained on, since every model can do that without a lot of effort. The idea is that the model has good generalization, so it can "understand" the data to the extent this is possible for a Machine Learning model. This understanding, which is often referred to as generalization, enables the model to perform well in all kinds of datasets that are somewhat similar to the training dataset. So, if it has performance P on the training set, it is expected to have a performance comparable to P in whatever testing set it is applied on. Usually the latter is a bit lower than P, but it is close enough to render the model robust.

What's more, optimizing a model involves going through several parameter combinations in an attempt to maximize or minimize a performance metric. Naturally, it's not possible to cover the whole parameter space but that's a compromise you need to make if you are to ever finish optimizing a model. Besides, the number of parameters in a model can be relatively high, making the possible combinations too many to go through one by one. Still, you can pick diverse enough values and calculate the performance of the model for each one of the combinations. It's best to stick to a particular metric for measuring the performance, so that you can also compare the model with other ones you may try out later.

Furthermore, the sampling used when testing a model, to ensure that it is robust enough, needs to be random or close to random. Ideally, it would be part of a K-fold cross-validation process, so that it is trained and tested on mutually exclusive samples of the dataset. Naturally, the parameters used in each one of these K folds are going to be the same. Also, higher values of K, such as 10, tend to be preferable as they allow for a more accurate exploration of the model's performance and its consistency. However, for larger datasets smaller values of K are often used, such as 3. Of course, you can achieve the same result as K-fold cross-validation if you get multiple random splits of training and testing, without them being mutually exclusive. Yet, this strategy may require more samples to be taken and is therefore more time-consuming.

Finally, optimizing a model is always data dependent, since an optimal model for one problem is bound to be sub-optimal for another, different, problem. Even though there is such a thing as transfer learning, which is basically using a model trained on a dataset on a somewhat different dataset, with some success; this is more of an exception to the rule. The reality is that for most Machine Learning models, you need to be mindful about where you apply a trained model, if you want it to provide reliable results. However, if the two datasets are of the same domain, such as texts from the same genre of books, it is possible to use your model more broadly.

Summary

- Over-fitting in a model is a state whereby the model is overly dependent on the training data and unable to predict anything else reliably. Over-fitting is a characteristic of an overly complex model and it's the most common issue in model training.

- The bias-variance trade-off is an important law describing a model's performance. It claims that for a given model, the bias and the variance of that model are reversely proportional. This law is important for understanding and optimizing the model, for a given problem.

- Machine Learning and Statistics are quite different as fields and ought to be differentiated when learning about them. This can help you understand them better and also know when it's best to use each for optimal performance in the ensuing models.

- Statistical models can be easily combined with Machine Learning ones, for even better results. A classical case is when a specialized statistical model is used first to figure out which features are the best ones and then use a Machine Learning model based on these features only.

- Heuristics are very important in Machine Learning and Julia offers a unique advantage as a high-level programming language when it comes to developing and using them.

- Ensembles in Machine Learning are particularly important although they are not a panacea since they too can over-fit or under-perform. Ensembles can be homogeneous (i.e. comprising of the same type of model) or heterogeneous (i.e. comprising of different types of models), and they tend to use a heuristic for combining the results of their component models.

- When optimizing a Machine Learning model, be it a basic one or an ensemble, it is important to bear in mind that it's the combination of parameter values that needs to be optimized, in reference to a given performance metric, such as accuracy rate or mean square error, for classification and regression respectively. Also, all models are subject to over-fitting and under-fitting, so proper testing need to be done, such as through K-fold cross-validation.

Questions

1. How can you improve the performance of a model in terms of bias?

2. How can you improve the performance of a model in terms of variance?

3. How can you improve the performance of a model in terms of both bias and variance?

4. Some people claim that ensemble models are better because they don't over-fit. Is there any truth in that statement?

5. What's better to have in a predictive model, higher bias or higher variance? Explain.

6. How can a heuristic improve a Machine Learning model used for predictive analytics?

7. Are heuristics found in Machine Learning only? Give a couple of examples or counter-examples.

8. How are AI-related models related to Statistics? Explain.

9. What's the best way to build an ensemble to have better performance overall?

10. Someone in your team has developed a Machine Learning model that has low bias and low variance for a particular dataset. Would you use it for your dataset, which has the same number of variables, but it's inherently different in all other aspects?

CHAPTER 10

Machine Learning's Impact

Although Machine Learning is an important component of the technical part of data science, when it comes to the real world, its value needs to be assessed and justified. As this laborious and often challenging task usually falls into the hands of the data scientist, it's important to know this topic well, and be able to convince others that a particular project will improve ROI, streamline a pipeline, or increase the quality of a product or service.

In this chapter, we'll view the importance of Machine Learning in an organization. We'll start with the rationale of Machine Learning when statistical models have an obvious advantage, how Machine Learning models have some levels of interpretability, and the projects where Machine Learning shines as a methodology. We'll continue with some ideas about how you can defend the more advanced Machine Learning models, especially those that are network-based, and how you can leverage the interpretability-performance trade-off effectively.

Why Machine Learning when statistical models are more transparent?

The first question a project stakeholder may ask you when you come up with a Machine Learning-based solution is why bother? Statistical models were fine and as a bonus, very transparent and easy to explain. Even those less technical could make sense out of them, since they had taken a Stats class at one point.

To answer this question, we need to define Machine Learning and how it is different from Statistics, emphasizing that it's a very broad spectrum of models. So, if transparency is valued, mention that there are transparent Machine Learning models available. At the same time, you can combine any Machine Learning model with a statistical one in an ensemble setting. Also, it's important to note that a Machine Learning model is bound to have better performance, something that can translate to lower costs due to erroneous predictions. If all else fails, you can use the argument that the company's competitors are already using some form of Machine Learning in their pipelines.

It's also important to be convinced yourself that Machine Learning models add value to your organization, and for you to have a working knowledge of many of them. If you only know about ANNs, for example, this may not be enough. Also, it's important to be able to come up with your own solutions, perhaps combining different approaches. If you can show that with a custom-made heuristic, you can design a Machine Learning-based solution which would be unfeasible with conventional approaches, you can sway the stakeholders' view on the matter and secure a proof-of-concept at the very least.

Furthermore, if you have enough time, it may be worthwhile to run some experiments on the available data using both a Machine Learning model and a statistical one, to compare their performance. This way, you may gather sufficient evidence to make the case that a Machine Learning approach is worth it, even if, in many cases, the transparency of the statistical models is sacrificed to make this possible.

Accessible levels of interpretability

Even if Machine Learning models are not transparent in general, there are certain levels of interpretability in such a model. For starters, even if most people don't realize it, the confidence score that often accompanies a prediction is handy information. Although this is often interpreted as a probability, it is, in essence, a heuristic that depicts how sure the system is about every one of its predictions. Note however, that this is usually done for classification problems since regression confidence is a bit more challenging to assess—some would say impossible. This confidence metric may not seem like a lot. Still, the fact that there is a popular evaluation metric called Area Under Curve (AUC) in binary classification, demonstrates its usefulness. Also, if a particular Machine Learning model doesn't yield a confidence score, it can be modified so that it does, by making use of the data used internally in its decision engine.

What about other aspects of interpretability, such as how each feature contributes to each decision made and how features rank overall? Such aspects are also available in Machine Learning, though they are limited to a certain kind of method, namely those related to decision trees. Decision trees and ensemble models based on decision trees have direct access to a systematic use of each feature as the tree makes its every decision. It's easy to trace each decision to specific choices the tree has made, each choice utilizing a feature in the dataset, though sometimes a feature is used more than once. Of course, you don't obtain any weights for the feature used, when utilizing a decision tree, but you can figure out how important each feature is based on how often it's used and if it's used closer to the beginning of the prediction process.

When it comes to decision tree-based ensembles, things are a bit less descriptive. A random forest, for example, will canvass all the trees it utilizes and counts how often each feature is used. Then it will provide a relative frequency metric that will be used to rank the features in terms of importance—the more important a feature, the more trees will be using it and in some cases use it more than once. Naturally, this ranking is specific to that particular model, even if some people use it to assess features in general, a strategy of debatable effectiveness. However, it does show that tree-based ensemble models are interpretable to some extent.

As for models that are black boxes by design, there is something you can do there too, though it's not super easy and not always accurate. Namely, you can find a transparent model, train it on the same data, and observe how well its predictions correlate with the black box model you are trying to interpret. If the predictions of the two models correlate strongly, such as over a predefined threshold value like 0.9, you can use the transparent model as a proxy for the black box one. So if your black box model is similar to a decision tree model in its predictions, you can interpret the decision tree model and use that interpretation for your black box model.

Whatever the case, even black box models tend to have an estimate of how certain they are about their predictions, something referred to as confidence. Although confidence is commonplace in statistical classification models, many Machine Learning models exhibit it too, which enables them to be gauged more thoroughly through specialized metrics like Area Under Curve (AUC), used in binary classification problems. Confidence also aids the high-level understanding of a model as it's something that every project stakeholder can comprehend and appreciate.

Domains and projects where Machine Learning shines

The No-Free-Lunch theorem states that no model can yield the best performance across all possible problems. In other words, if you see a performance edge in a certain model for a particular problem, it is bound to perform worse in other problems, since there is no free lunch when it comes to data models. That's why it's important to know for which domains and for which projects Machine Learning models work well and are therefore the optimal choice.

Such domains include industry sectors that deal with fast-moving data, usually complex, and which is analyzed for internal purposes. For example, the telecommunications industry, as well as the IT sector, benefit a lot from Machine Learning models. What's more, in areas like manufacturing, transportation, logistics, and engineering (e.g. failure identification and prediction), Machine Learning thrives. Web-based industries are, of course, the domain that utilizes Machine Learning the most, particularly when it comes to modern applications dealing with vast amounts of data, such as search engines and social media.

Also, certain projects across different domains lend themselves to Machine Learning, such as self-driving cars and other robotics applications. The high performance and effectiveness they apply to processing sensor data is incredible. Also, any noticeable lag in these applications could be catastrophic for the whole system, so all the data available needs to be analyzed fast. Chatbots are no different. Even if it is possible to make a chatbot without the use of a Machine Learning system, such a bot would not sound natural while its scope of responses would be greatly limited. Nowadays, chatbots involve more than just Machine Learning to make them more realistic, but the core engine behind the analysis of the language remains Machine Learning-based.

In general, any domain or project that requires particularly high accuracy in the predictions, deals with unrefined and usually noisy data, and doesn't require easy interpretability, is suitable for Machine Learning.

Defending a black box model

But how can you defend a black box model if that's what you need to go with to ensure the performance required? Defending such a model is not an easy task, but it's feasible, to some extent, especially if you have already convinced the project stakeholders that performance is more important than transparency, for that particular project.

First of all, you need to have a solid understanding of the problem at hand. Of course, it's easier to bypass all this and just plug the data into the model, but the real world is quite different from Kaggle competitions. Understanding the problem well and the data involved enables you to have a common frame of reference with the project stakeholders and communicate effectively. If you understand how the data relates to what you are trying to predict, the model's opaqueness won't seem that bad. You will still be able to explain the mapping performed by the model on a high level, even if the model won't be able to.

What's more, you need to comprehend the black box model you are using, enough to be able to explain it in layman's terms. This way, all the uncomfortable sense of mystery of the black box model may cease being so problematic because the project stakeholders understand that it's just a complex mathematical model. Of course, this won't make the model less opaque, but it will be perceived as less of a liability, at least for the length of the project. Also, being able to explain the model in terms everyone can understand can build trust in that you know it well enough to manage it reliably.

In addition, you can separate the insights part from the predictions part in your project and ensure that the black box model is used only in the latter. After all, you don't need a fancy Machine Learning model to do exploratory data analysis, which is a good source of insights. Also, you run a variety of models in your data before ending up with a black box one, gaining a lot of useful information from all of them. This way, when you start using the black box model, you can focus on a particular task, which, although difficult to interpret, may still add value to the project, complementary to the insights you've already found.

Moreover, you can dig into the requirements of the project and pinpoint which one has to do with performance. You can emphasize performance improvement and how the black box model makes it possible, at the small cost of some transparency. If necessary, you can run some experiments to show how this model, despite its opaqueness, outperforms any existing model significantly. This way, you can justify the use of the black box model, even if it's not easy to understand its reasoning.

Finally, you can always say that the black box model you've built is a temporary solution and that as the technology evolves, you'll be able to replace it with a transparent model that yields comparable performance. After all, it's a matter of time before transparent, high-performance models are a viable option, thanks to the necessity of transparency in many situations. This way, even if the project stakeholders are not particularly happy with the black box model solution, they can put up with it if they know it's going to be around for a short time.

Data is important!

What many people in the business world often neglect is that the success of a model usually depends more on the data and less on anything else. If your data is of bad quality, this will carry to the predictions made from this data— Garbage In, Garbage Out, or GIGO for short. However, we ought to discern between bad quality data, which is generally useless, and slightly noisy data, which is often treatable through data engineering.

Noisy data is common but it's not a huge issue as long as it contains a strong enough signal. Most of the advanced Machine Learning systems can even take care of all this themselves, given that it's not overwhelming. Bad data, however, is data where the signal is either very weak or non-existent, making even the most advanced system incapable of doing anything useful. Instead, these systems get hang up on the random patterns of the noise interpreting as a signal, leading to over-fitting. The quality of the data can be decently gauged in the data exploration stage.

Note that the amount of data is also very important. This is due to the fact that more data usually equals more information and stronger signals. Also, it has been observed that all predictive models tend to perform better as the amount of data used to train them grows. What's more, AI-based systems, in particular, tend to harness this additional data very well, giving them a clear edge in performance. The latter is often significantly better than specialized people who are domain experts in these predictive tasks, for instance, medical doctors diagnosing a disease. That's why data science and Machine Learning are so popular in this era of big data, where there is so much available data.

It's also important to remember that although having lots of data is great, it's even better if at least a portion of it is curated. That is, it has labels related to what we are trying to predict. Even though there are models that can work with unlabeled data and make sense of it, they are not as useful, in terms of value added to an organization, as their predictive counterparts. The latter, however, require some ground truth to start with, before they can make their predictions, and the more of that ground truth we have, the better the models. Curating data may be easy at times, but usually, it carries a cost, which is why it's not as easy to come by. Fortunately, there is an advanced Machine Learning methodology called active learning, which enables us to curate data faster through a collaboration of a computer and a human, but even this methodology requires some labeled data to get started.

Finally, the variety of data, translated into a diversity of variables, is significant too. After all, when developing a Machine Learning model, we don't just opt for performance but also stability, something that is made possible through diversity in the training data. High performance will impress the project stakeholders and help the project get started, but stability is what is going to convince them that it's truly valuable and help the project for future iterations.

Leveraging the interpretability-performance trade-off

It's really challenging to combine interpretability and performance in a model, as there is a trade-off between the two. Even though this trade-off is not absolute, it's still there for the vast majority of cases. That's why it's important to be familiar with it and know how to use it to maximize the value of any given data science project. Note that intermediate options exist, where there is some interpretability and some increased performance—there are options that are more interpretable (with less performance) and others that yield better performance (but are black boxes). Deciding among these options in a way that makes sense for the project stakeholders is not a trivial matter as it requires a consensus that's often challenging.

What's more, performance doesn't grow linearly, especially after a certain high level of performance is reached, also known as the Pareto principle. That's why in cases like this, it may make sense to settle for a less performing model that's also somewhat interpretable, such as a boosted trees classifier, rather than the black box model that's slightly better performing, such as a deep learning classifier. It's not an easy call to make, but it's something worth considering, especially when a small performance gain is not essential for the project at hand.

In addition, interpretability may not be required all the time, so in many cases, we may need to trade it, or at least some of it, with a boost in performance. For example, once the PoC of a data science project is complete, the stakeholders may realize that the model's performance could be improved and that this increase will translate into better revenue. So, the transparency of the model may not be as important in such a scenario, leading to a more opaque model being the favorable solution.

Finally, the interpretability-performance trade-off is not set in stone, since there is the potential of models that combine both. Naturally, this goes against our intuition, in a way, but then again, lots of things happen that are counter-intuitive. Entropy, for example, may increase over time for the universe yet decrease in certain hot-spots, at least for a given time frame. Similarly, it is possible for the interpretability-performance trade-off to be bypassed in certain peculiar cases. Identifying these cases is not easy, but it's what adds a lot of value in a data science project since it is more likely to yield a consensus of contentment, something quite rare in any project.

Useful considerations

Machine Learning is not a panacea when it comes to data-driven problems in a business. Just because it offers a variety of nifty data models and data processing techniques, doesn't make it a cure-all. In a business problem, the data science model is just one of the many factors that need to be addressed. Even the perfect data model may be insufficient if it's not shown to be relevant, and the results it yields are not used properly. Also, the comprehension gap between the project stakeholders and this technology needs to be bridged, and a good argument needs to be made regarding the value-added through its usage. Otherwise, it may be seen as a potential liability that's best avoided.

This ties in well with another useful consideration, namely the fact that Machine Learning know-how is not a substitute for other qualities of a data science professional. After all, many people can train themselves to handle Machine Learning models, but a good data scientist goes way beyond that. Such a professional can understand the requirements of a project, liaise with the business, collaborate with the other professionals involved, and process the data so that it's more suitable for the model. Also, a data scientist will be able to take responsibility for his work, especially any privacy-related concerns, explain his work to the extent it's possible, and maintain the models created as new data becomes available.

In addition, ethics is particularly important in data analytics work, especially when dealing with complex data. Besides, Machine Learning may do wonders in the analysis of data, but it's not sufficient for facilitating decisions *about* the data. Things like whether certain fields should be used or not, for example, because they may be linked to PII. These matters have more to do with one's ethical status rather than his Machine Learning aptitude, and in many cases, such a factor is more important. There is a reason why data science work is undertaken by people, even though a large part of it can be automated.

Moreover, understanding the data and how your insights could be used is as important as the process of developing those insights. After all, as a Machine Learning professional, you are part of a group, not a solo worker, even if you work as a contractor. So, caring about how your work is utilized and how it can be improved, based on the feedback you may receive from stakeholders, is crucial. However, sometimes you may need to take action to obtain this feedback and figure out how your work fits in the bigger picture. At the very least, this can help you hone your business skills more.

Furthermore, there is no "one size fits all" in data models so it's important to be familiar with various Machine Learning techniques. It doesn't take an expert to know that some data models work better than others, such as RNNs for time series analysis and CNNs for image, sound, text, and video data. So, if you are fixated on a certain kind of Machine Learning model, you may not make the most of your data. Also, it's often the case that a combination of Machine Learning models, called a model ensemble, is preferable due to the additional performance it yields and that it's generally more robust than every one of its component models.

Finally, just because an organization wants to go with a Machine Learning solution because it's considered state-of-the-art, doesn't mean that it can or that it should. This may look like you are undermining yourself, but a project that doesn't have the data for a Machine Learning-based analysis is better off not happening. Besides, there is always the potential of undertaking such a project in the future, once the required data is in place. Rushing into it just because others are doing so is a recipe for failure. After all, if you are knowledgeable enough about data science, there are plenty of ways you can add value to the organization beyond Machine Learning. This is one of the key topics explored in the book "Data Scientists Bedside Manner" published recently by Technics Publications.

Summary

- The argument why Machine Learning models are worth considering even if statistical models already deployed provide value, is not an easy one but it is necessary, in order to convince the project stakeholders about the potential value Machine Learning can bring to the organization.

- The performance of Machine Learning models is usually the deciding factor for using them, yet this often needs to be demonstrated.

- Interpretability in a Machine Learning model exists to some extent, though it's not usually as pronounced as that of statistical models. However, decision trees and ensemble models based on them tend to be quite interpretable.

- Machine Learning is particularly good for certain domains such as telecommunications, manufacturing, and web-based technologies. Also, projects related to robotics tend to rely on Machine Learning methods, while any area that requires high-performance predictive models can benefit from Machine Learning.

- You can defend a black box model in various ways, the most important of which is having a solid understanding of the data at hand and the problem you are trying to solve. Also, knowing your model well enough to explain it in simple terms, as well as being able to provide useful insights, can also be very useful strategies.

- Data is of key importance when creating a Machine Learning (or even a statistical) model since it is what contains the signal, which is used to understand the underlying information and make predictions based on that information. The key aspects of data that you need to have in mind is its quality and quantity, both of which need to be high to maximize your chances of a well-performing model.

- There is usually a trade-off between interpretability and performance in a model. Namely, the larger the performance, the less interpretable the model. In some cases, one or the other is more desirable, which is why it's important to look at a project's requirements rather than the performance factor alone.

- When applying Machine Learning in the business world, it's important to remember that Machine Learning, albeit powerful, is not a panacea since there are other factors that need to be considered. One such factor is the qualities of a data professional, especially things like ethics, something particularly useful in cases where private data is concerned. Also, it's important to consider other options beyond Machine Learning at times, since versatility and an open mind are definitely value-bearing qualities in a data science project.

Questions

1. Can every statistical model be replaced by a Machine Learning one and still add value to an organization?

2. What constitutes as interpretability and why is it important from a business perspective?

3. Would you use a Machine Learning model instead of a statistical to predict the exact demand in energy usage for a factory?

4. What kind of a Machine Learning model would you use to tackle a marketing-related problem that is linked to the strategy of a company?

5. Could you combine a statistical model with a Machine Learning one for better results?

6. What would you say to a project stakeholder who is convinced that Machine Learning is just a modern way of referring to statistical models?

7. Can you explain to a business person how a Machine Learning model functions, through the use of statistical knowledge only?

8. What's the value of confidence in a Machine Learning model's predictions, when it comes to the application of that model in a business setting?

9. How can you assess the quality of data that you are required to use in a predictive analytics model, for example?

10. Are there any statistical models that don't have Machine Learning counterparts, for solving business-related problems?

CHAPTER 11

Additional Thoughts on Machine Learning and Julia

Let's now take a look at some useful considerations regarding Machine Learning and Julia overall. After covering some various high-level considerations, let's now cover some additional topics that are relevant in a data science project and go beyond Statistics. We'll look at Machine Learning and Julia in general, focusing on aspects of them that are very important although they don't get enough attention. Namely we'll look at Machine Learning as constant work in progress and how transparent Machine Learning is a real possibility, regardless of the transparent AI case, which we'll talk about briefly in the next chapter. Also, we'll explore how insulating yourself from all the Machine Learning-related noise can be beneficial, particularly when delving deeper into Machine Learning through reliable channels. Next, we'll take a look at how we could reconcile the Machine Learning and Statistics dichotomy and look at the two fields as a whole. Finally, we'll look into a couple of Julia-related topics, namely how you can use Julia as a platform for innovation and how you can learn it in more depth, mostly to access this innovation potential.

Machine Learning as work in progress

First of all, let's look at Machine Learning not just as a respectable sub-field of data analytics, which is also at the core of data science, but as something that's never settled completely. Unlike Statistics that seems to have solidified overall, Machine Learning is nowhere near that stage of evolution.

For starters, Machine Learning is a fairly young field, and as a result, there are bound to be new discoveries as well as improvements on existing methods. After all, most people who embark on learning Machine Learning do so with the enthusiasm of a pioneer, something that captures the sentiment toward Machine Learning nowadays. Even if the days of groundbreaking innovations are behind us, there is still a lot of progress in the field, making it even more interesting and promising.

In addition, there is a lot of emphasis on Machine Learning, so many researchers prefer to focus on this field instead of some other, more established aspects of data science. As a result, there is more potential for scientific innovation in Machine Learning, since it's usually funded researchers that make the most progress in a field. Even if much of this research is not initially known, eventually it makes it to the users of Machine Learning methodologies. Especially today, when there are a plethora of Machine Learning-related startups—the gap between innovation and the application of it is getting narrower and narrower.

What's more, Machine Learning taps on a much wider approach to data analytics, void of predefined mathematical models of the data at hand, so there are plenty more possibilities. Many of these possibilities rely on heuristics, which play an important role in Machine Learning and which are in great supply too for the more imaginative researchers. As a result, Machine Learning can grow in various directions, given there is someone to lead the exploration, something that's becoming increasingly easier as the field grows. The fact that developing and implementing new heuristics these days is easier than ever helps all that immensely.

Furthermore, the people involved in Machine Learning tend to care more about innovation, even if many of them are not researchers in the field. After all, with the programming aspect of Machine Learning having become easier and more accessible to everyone, particularly with languages like Julia, it's fairly easy to tinker with Machine Learning models and experiment with new ones. This can bring about a new wave of models that are developed by users rather than professional researchers, though it's more likely that these users will have read a few research papers on the topic first.

Finally, Machine Learning incorporates aspects of AI, such as the ANN-based models, making it much larger and much more promising as a field. ANN models have experienced a renaissance lately as the technology has evolved enough to make these computationally heavy models a viable option. As a result, there is plenty of research in this area, and their incorporation in Machine Learning makes the application aspect of these AI models as easy as pie. In fact, these models are so widespread due to the high performance they yield that they have become almost synonymous with Machine Learning for many people, especially the newer waves of practitioners.

Transparent Machine Learning

Although there are already a few methods under the Machine Learning umbrella that are considered transparent, mostly decision trees, and to some extent random forests and boosted trees, transparent Machine Learning is still not a full-blown reality yet. The idea is to be able to enhance the existing Machine Learning method and possibly create new ones too, that provide a rationale of their decisions as well as a confidence metric for their predictions, in the case of predictive analytics. Such a task would not only help the data scientists using these methods but also all the stakeholders of the corresponding projects. After all, transparency in predictive analytics is often a requirement in data science work.

Also, transparency is something essential when it comes to GDPR since there is a requirement in this legislation that implies interpretability. Although GDPR is not strictly enforced, at least not at the time of this writing, it is clear that black box models don't look good in the eyes of the community. As a result, it's more likely that in the future transparent models may be a requirement in certain domains, particularly those dealing with crucial matters in the society. Once this happens, it's only a matter of time before other countries, not affected by GDPR, will follow suit to increase their chances of EU countries doing business with them.

What's more, the idea of transparency already manifests in statistical models. After all, they are designed this way, and the fact that they are easy to interpret plays an important role in their popularity and establishment in data science. Sometimes they are preferred even though better performing models are available, due to the transparency they offer. Besides, transparency equals lower risk in the minds of the project stakeholders, a perception that's usually quite hard to shake. So, overall transparency is not some pipe dream but a reality already in some form. We just need to extend this reality to include Machine Learning too.

In addition, transparency in Machine Learning can open new avenues of research and help us devise new approaches to data analytics, since to achieve transparency, we need to go beyond the conventional thinking that characterizes Machine Learning today. This leap of perception and understanding can help us look at things from a different angle and capture the essence of predictive analytics. Then, it's only a matter of time before this can bring about new ideas and new models that can disrupt the data science field, making it even more valuable to the world.

Moreover, transparent Machine Learning is quite possible given the right heuristics and a more creative approach to the problem at hand. However, this would require a different approach to the matter, letting go of many assumptions about how things should be, remnants of the older way of thinking about data analytics in general. Nevertheless, it is possible, and it's a matter of time before enough Machine Learning experts come forward with this alternative approach to rendering a paradigm shift in the field through the use of these heuristics and their incorporation in data models.

Furthermore, transparency in Machine Learning is all about understanding the features-to-target-variable mapping ourselves—otherwise expecting this from the data model would be next to impossible. The idea of AI becoming self-aware and solving all of our problems for us may be attractive, but it's not realistic in any way. Even if such an AI would come about, it's doubtful that its first priority would be to explain everything to us in ways we can understand. That's why if transparent Machine Learning is to become a reality, we need to be the first ones to make sense of the mapping between the features and target variable.

Finally, transparent Machine Learning is something that requires innovative effort from our part as it's unlikely to come about from conventional frameworks. The latter are good at what they do, but they rarely go beyond their own assumptions about the data and how the predictions should take place. For a more transparent approach to be possible, we need to go beyond this framework and come up with a novel framework that is transparent by design. Afterward, we can look into how this can be integrated with the existing frameworks if it's required or adds value.

Removing the ML-related noise

Although it goes without saying, it's important to have a discerning eye when it comes to Machine Learning content encountered on the internet. Unfortunately, much of what's out there is not reliable material you can use to educate yourself on this subject. Even if it is theoretically possible to grasp the fundamentals of Machine Learning in a few months and become comfortable with it during that time frame, the majority of people would require a lot of effort and guidance to do that. So, if a video or crash course promises to make you an expert in a short period, most likely, it just wants to take advantage of those lacking the patience to undertake this task properly.

What's more, many Machine Learning-related articles out are often written to promote a certain agenda, which creates unconscious biases in their readers. Always check out the authors of these articles before reading them and make sure that their intention is to inform and educate. Some of them just want to get more business for the companies they work for, or for themselves if they are solo entrepreneurs.

Additionally, Machine Learning is a fairly new field, so the term "expert" is a bit relative. Just because some company has employed someone for Machine Learning tasks doesn't make that person an expert, while the true experts don't label themselves as such since they are aware of the vastness of this field and prefer to show some humility about their expertise.

Furthermore, it's important to first assess your own needs when it comes to Machine Learning. Most of the available content is for beginners, which is fine, but it may not suit you if you have experience. Also, much of this material is for Python users, something that may not interest you as much if you are into Julia or some other high-performance language.

All in all, it's best to exercise good judgment when it comes to figuring out what you are going to invest your time in when it comes to Machine Learning. Not everything is for you, and as you progress in this field, your needs are bound to evolve, too, something you need to always keep in mind. Besides, a lot of your learning comes from practice, particularly when it comes to new methods and approaches, something that also applies to refining your Julia know-how, as we'll see later on in this chapter.

Delving deeper into Machine Learning through reliable channels

Start your learning with reliable sources, such as legitimate books made available by a technical publisher who has provided some kind of quality check before these books are released to the public. Fortunately, there are plenty such books and even if some of them are not as good as others, overall, they are much more reliable than the stuff you find available for free.

Videos that follow the same quality standards are also a worthy alternative. These are videos usually published by reliable platforms, as well as videos of conference talks and workshops on this subject. The PebbleU site (www.pebbleu.com) is a great example of this. Even though videos like that are, unfortunately, a minority compared to the alternative options, they are easily accessible and don't require a huge investment from the learners' part. In the long run, they save you time since they don't require you to go through a whole lot of them to find what you need. So, even if you don't find such a video particularly useful, it may still be useful to others, while it may also lead you to the videos you can benefit the most from.

The same goes with Machine Learning blogs, since that's where the largest issue lies. After all, many such blogs are managed by newcomers to the field or people who have no idea about Machine Learning and just gather articles they find, mostly from amateur content creators. However, there are quality blogs too that focus more on the facts than on the traffic their content may bring about. These may be rare, but they tend to have a clearer view of the topics they delve into and dare to take a stance on important matters, such as privacy.

Regarding courses on Machine Learning, these are also a tricky matter. Since courses are generally something their creators monetize heavily, there are many illegitimate courses. However, courses that are offered by a university or some educational platform, such as Thinkful, are solid, and even if they are not as accessible as the YouTube ones, they are definitely a worthy investment of your time and money. That's not to say that anything with a price tag is good when it comes to courses, but it's a useful predictor. The final decision, however, depends on other factors too, such as the course creators' expertise.

Finally, any script or package you find that adheres to certain quality standards, is a worthy option to consider too. After all, you can learn a great deal from functional code, particularly if it's in a high-level language like Julia. Not all Machine Learning scripts are educational, but the ones that make it into packages tend to be a bit better than others, plus they come with examples and documentation.

Reconciling the Machine Learning and statistical models dichotomy

Throughout this book, we've been looking at Machine Learning in a juxtaposition to statistical modeling. Although important for educational purposes, such a dichotomy may not be necessary or even meaningful, especially considering that a data scientist uses both, in practice.

When you reach that point, it will become clear to you that this dichotomy is not that rigid, actually. Most data scientists don't pay much attention to it and only consider it when they need to decide between a transparent model and a non-transparent one. Naturally, this choice usually comes from the project stakeholders.

However, for the dichotomy to go away completely and have a unified framework for Machine Learning and Statistics, we would need a somewhat different approach, possibly one that's based on heuristics. Such heuristics could be combined to form a possibilistic framework—not to be confused with the probabilistic one that has been done to death! Such a framework has already been applied successfully for the past five decades in AI, so there is definitely merit in the idea—even if current AI systems are more network-based than anything else. Yet even those systems were found to be compatible with the possibilistic framework as in the case of the ANFIS model, for example.

What's more, Machine Learning models can be combined with statistical ones through confidence, a powerful concept that's already used in both Machine Learning and statistical models. Confidence is usually modeled as a probability, but this doesn't have to be the case, just like energy in Physics can be measured in different ways, due to the fact that there are different types of energy in nature. If someone looks at things from a different angle and manages to disassociate confidence from probabilities, maybe the shift to a possibilistic framework is not such a huge leap after all. However, it's more likely that additional heuristics would be needed for this shift to be feasible.

Finally, the idea of unifying Machine Learning with Statistics is not new. It has been attempted before, although quite poorly, through the forcing of probabilities onto Machine Learning models. So, there is definitely merit in pursuing that, though it is highly doubtful that the probabilistic approach to this unification is the best way to go. Just because others have failed to accomplish this unification, doesn't make it any less possible. It just goes to show that an alternative approach is required.

Julia as a platform for innovation

So, what about Julia? How does it fit in this whole matter? Well, as we saw in the beginning of this book, Julia is ideal for prototyping due to its simplicity and high performance. As a result, it is well-suited for innovation, particularly in fields like Machine Learning, where there is plenty of potential for new methods and variants of existing ones. When it comes to creating new things, the sky is the limit, and Julia is like an aircraft in that sense!

What's more, Julia, on its own, is not a guarantee that you can innovate even if it makes it much easier to do so. You will also need some creativity and the courage to view things from a different angle. However, this is much easier than it seems, and it's usually the implementation part that slows many innovators down. For example, during my PhD years, I had developed a new approach for performing basic Statistics as well as correlation analysis in a way that the process was immune to outliers and other anomalous data points. The algorithms were sound, but it was very slow in the programming platform I was using at the time, which was MATLAB. Even after optimization, the corresponding scripts were so slow that for anything more than 1000 data points was pointless. However, when I applied the same algorithms years later in Julia, they were quite usable, and I was able to extend them to cover much larger datasets. All this would have been extremely difficult with alternative languages as I'd have to deal with the two-language problem. Plus, I am no coder, so it's unlikely that the implementation I would create would be the optimal one.

In addition, Julia enables you to create new data structures, extend existing ones, and generally work more creatively. Considering that a few years ago, most of the packages you see today didn't exist, it's truly amazing what was made possible through the language. The cherry on top of the cake was when the Knet package was developed, proving that even something as sophisticated as a deep learning framework could be done elegantly in Julia, all while making it a high-performing option for deep learning. Who knows what else is possible by employing a methodical and persistent approach using Julia?

Finally, innovating with Julia may seem challenging at first, particularly as you are learning the ins and outs of the language. However, once you gain an in-depth understanding of the language, things start to change, and it becomes something natural. Even if you don't have a lot of new ideas, you may use Julia to implement existing algorithms in a better way and perform some tweaks to them. So, if you work with it long enough, programming in Julia may become second nature to you and innovating in it a more accessible task, possibly one you find enjoyable too.

Becoming a Julia expert

But how can you learn Julia in more depth? I'm not talking about learning more features and functions of the language, but about learning it deeper and making it an extension of your mind, just like any other useful tool at your disposal. This fairly ambitious task is quite feasible through a simple method: practicing regularly and challenging yourself.

All this may sound obvious, but nowadays the instinctive response of the majority of data scientists when they encounter a problem is to search for a ready-made solution in places like GitHub, instead of figuring out a solution on their own. This is justified to some extent by the fact that most data scientists still use low performing tools, which make any kind of non-traditional task quite challenging or unappealing unless you are an experienced coder. However, if you use Julia, this sort of task is much more accessible, feasible, and even insightful. This accounts to a large extent to the rapid growth of the code base of the language, since it was first released.

Moreover, learning Julia in-depth is not something you can easily do in a fixed time frame as it's more of a mentality shift, detached from specific objectives. In other words, it's not so much about doing project A or B using Julia as it is being able to undertake any project in that language. It is all about making it your tool of choice for programmatic or data science-related challenges. It's shifting from viewing it as a cool alternative to whatever language you are using to the go-to language for the foreseeable future. It's really at this point that you can learn the language in-depth and make the most of it in your Machine Learning projects.

Finally, to learn Julia, you must care for either solving a problem or creating something worthwhile. It's not a recreational endeavor in its core, even if it can have recreation as a side-effect. In essence, it's all about committing yourself to learn as much as you need, to bring about results, be it a solution or a new method. That's not something a single course, or even a book, can do for you since it takes more than learning the essentials of the language. However, once you are focused on the end result, such as the solution to that challenging problem or the implementation of a new algorithm, you forget about what you don't know and figure it out as you progress. This way, you learn more intuitively and without the constraints of a curriculum, thereby going into more depth, something that's quite rewarding on its own.

Summary

- Machine Learning is a field that's in constant work in progress, due to various reasons, such as the fact that it's relatively new in data analytics and the wealth of research interest in it, both from professional researchers and from passionate users.

- Transparency in Machine Learning is a real possibility and something that stakeholders of data science projects often require. At the same time, it's something that can help users of these models and the field in general as it can open new avenues of research and bring forward new and potentially better Machine Learning systems.

- It's important to have a discerning eye when it comes to Machine Learning material, especially on the web. Be particularly cautious of things offered for free, whether they are videos, books, and courses on the subject.

- Reconciling the Machine Learning and Statistical models dichotomy is doable, given that a unified framework can come about, borrowing elements from both Machine Learning and Stats. For this to happen, however, lots of changes would need to take place in both of these fields.

- Julia is quite suitable for innovation, particularly in fields like Machine Learning, due to its ease of use when it comes to prototyping and its high performance. To make the most of this, you need to apply your creativity to new algorithms or variants of existing ones, perform a series of experiments and come up with new methods that address the shortcomings of the existing ones.

- You can learn Julia in more depth through regular practice and challenging yourself. Additional know-how of the language can come in handy, but in-depth knowledge will come mostly by using the language to solve challenging problems and creating new methods, in a committed manner.

Questions

1. Why is Machine Learning a constant work in progress?

2. What's the biggest danger of consuming Machine Learning material these days?

3. What's the most important thing when it comes to learning Machine Learning, in order to avoid getting hustled by opportunists in this domain?

4. What's the deal with the dichotomy between Machine Learning and Statistics when it comes to data models?

5. Can Machine Learning and Statistics be combined in a unified framework, when it comes to predictive analytics?

6. Why is Julia such a good option for innovating in fields like Machine Learning?

CHAPTER 12

Future Trends

Although all of the material covered in this book is relevant and precise at the time of this writing, things in this area change rapidly. This is why being aware of future trends in Machine Learning as well as in Julia, is crucial if you are to remain relevant in this field. As a bonus, all this can help maintain your enthusiasm for Machine Learning and the Julia language for years to come.

In this chapter, we'll look at various trends, mostly on the Machine Learning front. We'll start with how Machine Learning is becoming the de facto approach to data science, the new methodologies that are becoming available and are under the Machine Learning umbrella, a brief view of the hybrid systems based on Machine Learning and Bayesian Statistics, and the rise of actually useful AI that's less risky, namely AI that we can understand and extend without too much guesswork. We'll continue with some thoughts about the evolution of Julia in the years to come, as well as Julia's offspring language, Gen, and the potential it holds for data science.

Machine Learning as the de facto approach to data science

Depending on who you ask, you may hear different things about the role of Machine Learning in data science. For example, there are some die-hard advocates of Statistics who would swear that Statistics-based methods are the future and that Machine Learning is just an extension of them, if not just a fad. However, based on the popularity of Machine Learning approaches, it is clear that the latter is both independent of Statistics and on the rise. In fact, we would go so far as to say that in the years to come, it is expected that Machine Learning methods would be the de facto approach to tackling data science problems, with Stats playing a supplementary role while in some cases be entirely absent. It's also possible that Statistics is viewed as a kind of Machine Learning.

Statistics is bound to remain useful since there are certain tasks that are geared toward this approach. However, when it comes to predictive analytics, it is evident that there are no statistical models that can compete with the more advanced Machine Learning ones. The fact that there is a lot of research on making AI-based methods more interpretable shows that the days of statistical inference models are numbered.

Machine Learning is definitely not a fad, even with all the hype. After all, anything futuristic nowadays seems to be irresistible, especially for people who grew up with visions of a technologically-better future. Nevertheless, much of this is speculative and plagued by a toxic amount of optimism. However, many people in the data science field tend to believe that Machine Learning is just a tool in all this, and it doesn't have to be synonymous with a fully automated workplace. After all, Machine Learning is a framework, not a complete solution, and it's up to us to use it responsibly and in a way that is inclusive of people, instead of a way to rob them of their work. Perhaps once this is realized by the various stakeholders of data science projects, we will accept Machine Learning as what it is: a powerful tool that facilitates data science work.

New methodologies under the Machine Learning umbrella

Lately, there are some new methodologies that are under the Machine Learning umbrella. For starters, Reinforcement Learning (RL) is a fairly new methodology that, although similar to supervised learning, it has a series of differentiating elements that make it a distinct methodology. Also, as we saw previously in chapter 4, this methodology is purely data-driven, so it makes sense to classify it as a Machine Learning methodology. The fact that it is linked to modern AI-based methods solidifies this taxonomy choice. Besides, RL links to AI-based models, and the latter isn't going to go away any time soon.

What's more, semi-supervised learning is another such methodology that is often grouped with other Machine Learning entities. This methodology has to do with the development of models that are trained on a combination of labeled and unlabeled data (note that labels are known values for the target variable). This process makes it easier for the data scientists involved since acquiring lots of labeled data is challenging due to its high costs in terms of time and money. What's more, having lots of labeled data inevitably incurs human biases in the corresponding predictive model. Note that semi-supervised learning is not that new as a methodology, but lately, it has gained a lot of ground, and it's more prominent as a Machine Learning entity, especially when combined with fairly new methods such as Generative Adversarial Networks (GANs).

There are also other methodologies under the Machine Learning umbrella, such as *active learning*. This methodology entails a human-machine collaboration for learning something in a methodical and efficient manner. Specifically, for predictions that are made with high uncertainty (i.e. low confidence), a human annotator facilitates the task by providing what he believes to be the correct answer. This way, the system learns both from the data and the predictions, optimizing its overall performance and reliability. Active learning is particularly useful for complex predictive analytics tasks such as the analysis of 3D cardiovascular images provided by a magnetic resonance machine.

As the field of AI advances and we have more predictive models available, it is expected that the Machine Learning field is going to expand as it gathers more relevant methodologies under its wing. Also, the aforementioned new methodologies are bound to become more widespread, particularly in the application areas they are designed for, as more organizations start utilizing them either for profit or for enhancing the user experience of their clients.

Machine Learning and Bayesian Statistics hybrid systems

An exciting trend in Machine Learning that's worth mentioning here is that of hybrid systems incorporating Bayesian Statistics. The latter is a somewhat controversial part of Stats that is based on beliefs, probabilities, and what's probably the most fundamental formula in this aspect of Mathematics: Bayes theorem.

It's borderline impossible to overestimate the importance of this kind of Stats, which, although a bit older than Frequentist Statistics, has been largely an alternative approach to data analytics, at least until now. As computing power became more available and an increasing amount of people became aware of the merits of this Statistical framework, Bayesian Statistics gained popularity experiencing a kind of Renaissance lately. Even in data science applications, it is held in high esteem these days, while its transparency and ease of visualization make it a very attractive tool.

Bayesian (predictive) models are characterized by the following equation:

$$P(theta|data) = P(data|theta) * P(theta) / P(data)$$

where

- *theta* is an array containing the model parameters.

- *data* is the dataset available to use (aka evidence).

- *P(data)* is more of a normalizing constant, since it's impossible to compute. However, as it stays the same across all Bayesian models, it doesn't matter all that much.

- *P(theta)* is the prior probability, namely what we believe the probability of a given set of model parameters is going to be. In practice, a parametrized distribution is used to specify the value of this probability.

- *P(data | theta)* is the probability that corresponds to the likelihood of data given a particular set of model parameters. The formula for calculating this is always specific to the model we use, and it's used for the validation of these models (higher likelihood is generally good).

- *P(theta | data)* is the posterior probability, namely a probability distribution over model parameters obtained from prior beliefs and data. This is what we seek to calculate when developing a Bayesian model.

Naturally, the formula above is the same as the classical Bayes theorem, applied to data models. This really goes to show the universality of this theorem and how it applies to much more than the probabilities of two events. In fact, if there is one formula from Statistics that's worth remembering, it's this one. As for the age-old debate about which Stats is better, you may want to ignore it since most competent data analysts, some of whom are statisticians by trade, embrace both of them. A good data scientist ought to do the same since each one of these frameworks has its own merits and use cases.

From this formula, it becomes clear that probabilities in the Bayesian sense are quite different from the probabilities in frequentist Statistics. The latter refers to past observations (which is why they are usually referred to as *likelihoods*) while the former has to do with our beliefs about the future, based on what's observable here right now. That's why when we talk about probabilistic

Machine Learning or Bayesian modeling in Machine Learning, we are expressing the use of probabilities in this sense combined with conventional Machine Learning techniques, particularly networks (Graph Analytics). Since Bayesian Stats are geared toward smaller datasets, while Machine Learning models are adept at handling larger ones, the combination of these two frameworks has merit as it brings together the strengths of both these approaches.

The rise of useful AI that's less risky

Many people have wasted hours talking about AI as of late, in a way that resembles a science-fiction script. At the same time, others claim that it's just a fancy methodology that's Statistics on steroids or the evolution of existing Machine Learning systems. However, beyond these extremes, there is a trend of AI that is actually useful and down-to-earth, yet at the same time somewhat futuristic. The best part is that it's less risky than conventional AI and potentially groundbreaking.

Namely, there is the possibility of an interpretable AI, a system that not only yields predictions but informs us of its confidence about them, even for regression problems (also known as eXplainable AI, or XAI for short). Additionally, it traces back these predictions to the original factors considered, and is comprehensible in its functionality throughout. Such a system is the exact opposite of a black box, and it is something that can help the data scientists involved in many ways. Naturally, since the risk of biased models is limited, such an AI would be far safer and useful.

In addition, a system like this can handle different kinds of problems, not just predictive analytics, with ease. Imagine such a system being as robust with clustering tasks or dimensionality reduction, all while maintaining transparency in its functionality and its results. This is something that could benefit everyone and be incorporated in a pipeline effortlessly because people would be able to learn from it and trust it more. After all, it is often the case that transparency and interpretability are requirements for a data model when it comes to the project stakeholders.

Such a system is bound to have elements of existing predictive analytics systems though it does not have to adhere to any of the existing model structures in a verbatim manner. However, for a transparent AI system to work, it needs to be part statistical and part Machine Learning, combining the best of both worlds. Most likely, it will have a network architecture, though the heuristics it makes use of are bound to be completely different from the ones most ANNs use. Also, such a system is most likely to be implemented in Julia, due to the various advantages the language offers, as we explored in the first chapters of this book.

Naturally, such a system is not something that you would expect to come about as an evolution of the existing AI systems, even if some people are actively pursuing this strategy. However, once enough open-minded people attempt it, it is bound to happen and provide an interesting alternative to the black box systems that we are used to seeing in Machine Learning. In fact, it wouldn't be far-fetched to say that such an approach to AI would be paradigm-changing, much like quantum physics has been changing the paradigm of physics for the past few decades.

Julia's evolution

As you might expect, Julia is bound to continue evolving as a language and as a community. In fact, it would a big surprise if it doesn't since it has been gathering momentum even before version 1.0 was released. Naturally, ever since the production-ready version was made available, this pace has accelerated, while the amount of material available to newcomers to the language has increased dramatically. This includes both books made available by reputable publishers, professionally-made videos, and organized courses offered by universities as well as the Julia Computing company itself.

Also, we would expect Julia to become more of a household name as a programming language, to the extent that there are any such household names. After all, a few years ago, few people knew or cared about artificial intelligence, other than sci-fi fans and researchers. Yet, nowadays, everyone seems to have an opinion on the topic. Probably something similar is going to happen with Julia, too, especially if it continues to be marketed as an AI-related language.

In addition, we would expect Julia to become more widely used in different areas, beyond scientific computing. After all, Julia is a multi-purpose language, even if it was originally geared toward numeric computing and science-related projects. Perhaps as more and more people embrace this technology, we'll be seeing Julia being utilized in a larger variety of projects, such as Cybersecurity and robotic process automation (RPA). The creators of the language are doing their part in introducing it to different groups of people, and the whole community is becoming more diverse and inclusive.

Also, as there are more groups of users around the world, we would expect Julia to become more well-known among programmers and data scientists. After all, unless you are heavily vested in different technologies for this kind of work, you should be able to see the merit of Julia as a tool in these use cases. Besides, the Julia-related events such as JuliaCon are not a state secret while independent sites, including my data science and AI blog (Foxy Data Science), talk about the value of Julia as a programming and data science platform.

What's more, we would expect Julia to be taught in more areas, be it universities or online learning platforms. It doesn't take clairvoyance to see that this is a trend that's as real as the language itself. As more and more people realize that Julia is easier than most programming languages, without depending on vectorization for performance, it's a no-brainer to deduce that it's going to catch on in the educational world too. Of course, it may take a few years as it's not that easy to revise a curriculum in a university course. Yet, Julia Computing has already started giving courses on hot topics using Julia, while there is an abundance of tutorials on the language, even if they are mostly introductory. This, however, is bound to change as people become more aware of what Julia has to offer and start seeking it out as a skill.

Furthermore, we would expect Julia to be part of certain operating systems or at least specialized devices such as single-board computers geared toward certain computationally heavy tasks. After all, if computing resources are limited, it makes sense to have a powerful programming tool to make the most of them. Since Julia fits that description, it makes sense to expect that to be part of the operating systems such devices have. And as parallelism is a real thing in Julia too, it makes even more sense to use this language to harness the computing power of many such devices, organized in an array.

Finally, we would expect Julia to extend its functionality to more niche applications through specialized packages. As it's used in more and more applications, there is going to be a critical mass of users interested in them to render the creation of specialized packages that are properly made and maintained. This may not happen soon, but it's a real possibility, considering the evolution of the language so far and the community growth over the past few years.

Julia's offspring language

Gen, Julia's recent offspring language, is something that, although groundbreaking, hasn't yet received the attention it deserves. Specifically, Gen is a probabilistic modeling and inference platform that involves a series of areas, such as Statistics, Machine Learning, Computer Vision, Cognitive Science, Robotics, Natural Language Processing (NLP), and Artificial Intelligence. Gen can provide an abstraction of the mathematical parts of the processes from these areas and let the user focus on the result and high-level data science processes.

Although Gen is designed to help newcomers enter the field of AI by taking care of the more challenging parts of the corresponding techniques, it will help more experienced AI professionals, too, through its abstraction. Also, the fact that MIT has developed it brings with it a lot of confidence. Note that the combination of this simplified framework with Julia yields benefits unfathomable previously, which is why this new language made the news and put both MIT and Julia in the spotlight. You can use Gen through Julia by installing and loading the corresponding package, Gen, through the package manager prompt:

```
pkg> add https://github.com/probcomp/Gen
```

If you have time, we recommend you check out the corresponding session from the Julia conference of 2019 in Baltimore for more information and a couple of examples: https://bit.ly/2RKXmzZ. For additional information about Gen in general, you can refer to its official website: https://bit.ly/2KeoPWC.

Useful considerations

Even if we can make reliable predictions to some extent, there are always going to be some elusive phenomena that we won't be able to predict. Some of these have huge implications and may even change our perspective drastically. These are referred to as black swans, and they can be found anywhere, from the environment to the business world to even the tech world. So, whenever considering future trends, it's important to remember that some of the most impactful things are beyond our predictive capabilities. That's why we need to take future trends with a pinch of salt, since these may not manifest as we believe, based on our current information and insight.

What's more, there is a trend that is hard to bin into this or the other category, namely the development of synthetic data based on a dataset, using some advanced process, usually AI-related. Although it doesn't sound like much, this is one of the best innovations in the data world in our era. Privacy issues have plagued the world for years, and nowadays, people are more protective of their private data. So, Personally Identifiable Information (PII) is bound to become scarcer or have strict regulations regarding its use, something that's already happening in places like California and the EU (in the latter case, through the GDPR legislation). Being able to create synthetic data that carries virtually the same signal as the original dataset can help alleviate this issue. Also, it can help create more diverse datasets and therefore eliminate potential biases in the existing models, such as those models involving image-based data, where people of color are often under-represented in the datasets.

In addition, it's important to discern between what is an actual trend and what's just a fabricated hype, particularly in the case of Machine Learning. The former case is a genuine phenomenon that can be observed independently and has a lasting impact because there is a root-cause behind it that even if it's elusive, it's real. The latter scenario is just a very effective marketing campaign or the result of some chaotic chain of events that brought it about, much like noise that presents itself as a signal, in a complex dataset. That's why it's best not to believe everything you read in a data science article or video, especially one that's been published on social media. Indeed it's best to do your research and come up with your conclusions about what constitutes a trend and what doesn't.

Moreover, the data-driven approach to analytics is something that's gaining more ground than people ever expected before, a trend that is bound to continue. As such, it could be classified as a trend, since it's a fairly new thing. Believe it or not, there are still many people who would swear that Statistics is the best way to model data and that everything else stems from Statistics. Nevertheless, more and more people are getting into AI. With the latest developments in AI that make it accessible to anyone who wishes to learn, it's doubtful that the assumption-ridden paradigm has any future.

Finally, the use of specialized hardware for advanced Machine Learning systems is a reality and a trend that's bound to carry on. Much like GPUs came about to facilitate graphics-related tasks for a computer, nowadays there are TPUs, and even whole hardware configurations geared toward AI systems (here is an example, from an unexpected source: https://bit.ly/2xKR8Jo). Who knows what other seemingly exotic hardware technology is going to come about in the years to come as a way to make Machine Learning methods more efficient, secure, and reliable.

Summary

- Although there isn't a consensus on the topic of Machine Learning as the de facto approach to data science in the years to come, based on the latest trends, it seems to be the most plausible future.

- There are a few new methodologies that are now becoming part of what is considered Machine Learning: reinforcement learning, semi-supervised learning, and active learning. Reinforcement learning involves helping a system learn through rewards given whenever it performs well in a usually dynamic environment. Semi-supervised learning has to do with the use of both labeled and unlabeled data in a predictive analytics scenario. As for active learning, it's a methodology that involves the use of a human annotator working in tandem with a machine, for improved performance.

- Bayesian Statistics is a somewhat controversial part of Stats that is based on beliefs, probabilities, and the Bayes theorem. It's handy in and of itself, particularly in problems where there is limited information or high enough complexity, rendering conventional (frequentist) Statistics ineffective.

- Bayesian Statistics can be used in tandem with Machine Learning to form a hybrid approach to data analytics.

- Interpretable AI, although not implemented yet, is a real possibility for a new kind of AI that is more useful and less risky than conventional black box AI systems.

- Julia is bound to evolve in the years to come, with the most notable trend for it being education and general awareness about it becoming more widespread and diverse. Julia is bound to become richer in terms of packages and become more widely applicable.

- Gen is an interesting programming language stemming from Julia (it actually exists as a Julia package). The idea is to have a framework that provides an abstraction for the various models used, be it statistical,

Machine Learning related, or specialized, such as those related to computer vision. This abstraction can shield its users from the heavy math that often accompanies these models and enable them to work with them from a higher level standpoint. Gen was developed by a team at MIT and has been well-received by the general community.

- There are some useful considerations to have when it comes to trends in Machine Learning and Julia, such as the possibility of using synthetic data out of a dataset created through an advanced Machine Learning method (usually an AI-based one) as a way to ensure privacy, as well as the fact that the data-driven approach to data analytics is a well-established trend that's likely to stick around for the foreseeable future. Also, discerning between a real trend and hype is crucial when it comes to figuring out the potential future of Machine Learning, in particular.

- Although future trends are useful for remaining relevant as a data scientist, they are not always black-and-white as there are always highly unexpected things with disproportionally high impact happening (aka black swans).

Questions

1. Why is Machine Learning considered as the de facto approach to data science in the foreseeable future?

2. What are the most important new methodologies that are under the Machine Learning umbrella? Is this classification justified?

3. What's the key benefit of the hybrid systems that combine Machine Learning with Bayesian Statistics?

4. What is the actually useful AI that's on the rise these days, and how is it different from conventional AI? Also, how is this relevant to the real world (e.g. an organization that wants to utilize this technology in its pipelines)?

5. How is Julia expected to evolve in the years to come? How will this affect data scientists?

6. What's the deal with Gen, and why is it relevant to Julia and data science?

7. What's the single most important thing you can do to ensure that you are in the loop when it comes to Machine Learning and Julia, in the years to come?

Conclusions

We have finally finished our exploration of how Julia applies to Machine Learning projects in data science. Note that several other aspects of this subject have to do with your continued growth as a Machine Learning professional, whether you are a data engineer, data scientist, or businessperson. After all, the lines among these specializations tend to blur these days.

In this chapter, we'll cover how to remain relevant as a professional in this field. Next, we'll look at a few Machine Learning projects that you can do now in Julia for further practice on this subject. Then we'll share some good resources for learning more about Machine Learning as well as some further learning in the Julia language too. After that, we'll close this chapter with some final words about this topic.

Remaining relevant as a professional

Remaining relevant is particularly important nowadays where the competition has soared in this field. To remain relevant, you need to make sure that you are always updated about the latest developments in Machine Learning, continue gaining experience in this area, and improve your understanding of the field by going into more depth. The latter has to do with developing the mindset and intuition that enable you to view Machine Learning as a whole rather than the sum of disjointed parts. That's not an easy task, but it's something that can differentiate you from all the opportunists who enter the field for a quick buck or a flashy title on their online professional profile.

Remaining relevant is also related to walking the extra mile. Few people will ask you to learn something new or still experimental. However, taking that extra step that you are not obliged to take can open new doors, and new possibilities will present themselves sooner or later. After all, those who made it big in this field didn't do so by playing it safe and doing whatever everyone else had done. Perhaps that's why they gain such high rewards.

Moreover, remaining relevant is also about participating in events and communities (even online ones) that delve into this topic. If there isn't such a community in the city where you live, perhaps you can even start one. It doesn't have to be something extravagant. Sometimes a small group of people can be good enough plus given enough time, it may grow into something more. What's more, if this group is serious and committed enough, it can grow and lead to potentially making their endeavors, not just educational but also financially viable.

Finally, you can try to learn more about the business aspect of Machine Learning and AI since the two terms are used interchangeably in the business community. Understanding the technical aspects of it is a great foundation for leveraging it in projects. Still, it's often equally important to have a good grasp of what the business world thinks of it, managing the expectations of the stakeholders of your projects, and finding a way to smooth the communication between the two parties. It's not an easy task, but having this ability enables you to increase your value as an asset in the organization you work for and in the market in general. Tools and techniques in Machine Learning are in constant flux, as the methods evolve, but the ability to communicate with other people never goes stale.

Machine Learning projects in Julia

The best way to gain some experience in Machine Learning using Julia is to get down to some hands-on work using the language. Namely, you can take up projects either by yourself or with other people who are in the same boat, such as through a Meetup group. A good place to start is the various datasets available in the following sources:

- UCI Machine Learning Repository (https://bit.ly/34M1mFI)
- Kaggle (www.kaggle.com)
- Hiajin Wang's repository, from CMU (https://bit.ly/2RSMCiS)

Alternatively, you can create your own datasets to work with, by downloading a table or two from a database, after getting permission, and after getting rid of any PII, to avoid any potential privacy issues. This alternative would be more interesting, particularly if you can couple it with some domain knowledge related to the data you use. As a bonus, this will make the project much more relevant to you and rewarding in its own way, plus a project like this is bound to be more unique than one based on a benchmark dataset.

You can also create your own synthetic data, but that may not be as interesting a project, especially if you are new to this sort of thing. This involves generating random numbers on several dimensions and using them to make predictions. Naturally, the labels vector is bound to be somewhat structured, since if that were random too, the problem would be meaningless and difficult to solve. Synthetic data has been used to test drive new algorithms and explore their functionality in a controlled environment. Whatever the case, even though it's not as easy as it seems, it enables you to have as much data as you need and no unnecessary variables to dilute the signal in your dataset.

If you are comfortable with web scraping, you can also get data from the web, or whenever possible, through an API, such as Twitter. Naturally, some APIs will yield text data, but it's not too difficult to turn that into data that Machine Learning algorithms can handle. This, however, is a more specialized kind of know-how that goes beyond the scope of this book, and it's under the umbrella of Natural Language Processing (NLP). You can learn more about NLP from specialized resources such as videos and tutorials on the subject. Also, note that NLP is often considered a kind of AI since AI-based methods are usually employed in it, such as CNNs, RNNs and MLPs. Also, data deriving from NLP is a great use case of dimensionality reduction since the number of features involved in this sort of project ranks in the 1000s.

Finally, if you are more creative, you can come up with your own project from scratch. Nowadays, you can gather data even from a mobile device, be it through its sensors or via an app you deploy in the corresponding market. Data is very valuable, and if it comes from the real world, it is bound to be more information-rich. So, if you can come up with an idea of how you can leverage this to make something useful, you can have a very valuable project in your hands, which you can also add to your portfolio afterward. Ideally, you'd want to carry out projects like this by working with other people since they require a lot of effort and time!

Further learning on Machine Learning

You can always learn more about Machine Learning by exploring additional resources. The various videos available on certain learning platforms are a good place to start. Naturally, you can also take a course or two on this subject, so that you get a more in-depth perspective. Prof. Andrew Ng's Machine Learning course on Coursera is an excellent resource for this (https://bit.ly/3bniUu1), even if it is an introductory course. Other courses from various universities, as well as other sites, such as edX (https://www.edx.org), are also worth exploring. Also, bear in mind that although you can audit these courses for free, most such courses require a subscription, which is something you need to budget for, particularly if you plan to take several of them.

Further learning on Julia

Speaking of Julia, you can advance your learning of the language by using various resources. The best place would be the official documentation of the language, as well as the documentation and tutorials that accompany some of the packages. Additionally, you can check out specialized videos on the topic, particularly those developed by members of the Julia community. Most of them are available on YouTube at Julia's official channel (https://bit.ly/2XNM7dV).

Additionally, you can delve into the language through the JuliaCon conference that takes place annually (https://juliacon.org). Of course, traveling to the city it takes place may not always be feasible, but you can always check out videos of the various talks, which are made available after the conference finishes. All of these videos are published on the language's official channel. Also, you can participate in Julia hackathons as well as workshops, usually made available on one of the days of the conference.

Moreover, Julia Meetups are worth exploring, as well as certain sites dedicated to the language. One of the best is JuliaBloggers.com, which is an aggregator of various Julia posts from various technical blogs that delve into Julia programming. Not all of the posts will be relevant to you. Still, they can keep you in the loop regarding the latest developments, particularly when it comes to applications of the language, be it scientific or commercial.

Finally, you can explore Julia on your own and try to implement algorithms and metrics that don't exist in any of the packages right now—see Appendix C as an example to help you get inspired. Perhaps you can come up with your own methods, which even if they are not on par with the methods professional programmers create, they may be good enough to help you learn the language better and get more comfortable with its more subtle aspects. After all, it's like a tool, and in order to make the most out of any tool, you need to be able to see it as an extension of yourself.

Final words

Julia for Machine Learning has been a topic I always wanted to write a book about, ever since I learned enough about Julia. However, as it was a bit niche at the time, I decided to work on a broader topic, namely data science using Julia. Yet, even then, there was a lot of emphasis on Machine Learning, and although the language wasn't as developed as it is now, there were a lot of people working with Machine Learning models through the packages that existed at the time. Fortunately, things have evolved noticeably since then, and now Machine Learning is becoming a household name.

At the same time, many start-ups come about with Machine Learning as part of their business plan. The abundance of data nowadays in all kinds of domains makes this sort of endeavor more feasible and less demanding. Computational resources may still be an obstacle, but at least today you can outsource most of it to the cloud through cloud computing platforms, such as JuliaBox.

What's more, privacy matters are a major concern these days and oftentimes create additional barriers in data science projects. Particularly in Europe, where

GDPR has taken hold, it's very challenging to make use of customer data without their expressed permission. Otherwise, you risk getting a steep fine.

In any case, there is no doubt that it is a great time we live in right now when it comes to Machine Learning and the use of novel tools. Julia plays an important role in all this—as Machine Learning advances as a field, becoming more and more integrated with AI and other compatible fields, Julia is bound to evolve too.

Glossary

A

Accuracy (Rate): a commonly used metric for evaluating a classification system, across all of the classes it predicts. It denotes the proportion of data points predicted correctly. Good for balanced datasets, but inaccurate for many other cases. Accuracy is the default metric for the K-fold cross-validation method.

Activation function: see *transfer function*.

Active learning: a new Machine Learning methodology that makes use of a human annotator for assisting a machine in learning, by providing feedback to it for the most uncertain predictions it makes. Active learning is particularly useful for complex problems where high accuracy is a key requirement. Active learning is usually applied to AI-based models.

Algorithm: a step-by-step procedure for calculations and logical operations. In a Machine Learning setting, algorithms can be designed to facilitate Machine Learning and acquire knowledge by themselves, rather than relying on hard-coded rules. Algorithms in Machine Learning often rely on heuristics for their evaluations and decisions.

Application Programming Interface (API): an interface or communication protocol bridging different parts of a computer program or framework, simplifying the implementation and maintenance of any software involved. APIs are particularly useful in data science work.

Area Under Curve (AUC) metric: a metric for a binary classifier's performance, based on the ROC curve. It can take into account the confidence of the classifier when available and is generally considered a more robust performance index. AUC takes values between 0 and 1 (inclusive), and a higher value is better. Generally values around 0.5, or anything less than that, are considered bad.

Artificial Intelligence (AI): a field of computer science dealing with the emulation of human intelligence using computer systems and its applications on a variety of domains. The application of AI on data science is noteworthy and an important factor in the field, since the 2000s.

Artificial Neural Network (ANN): a graph-based artificial intelligence system, implementing the universal approximator idea. Although ANNs have started as a Machine Learning system, focusing on predictive analytics, they have expanded over the years to include a large variety of tasks. ANNs comprise a series of nodes called neurons, which are organized in layers. The first layer corresponds to all the inputs, the final layer to all the outputs, and the intermediary layers to a series of meta-features the ANN creates, each having a corresponding weight. ANNs are stochastic, so every time they are trained over a set of data, the weights are noticeably different.

Autoencoder: an artificial neural network system designed to represent codings in a very efficient manner. An autoencoder is a popular artificial intelligence system that is used for dimensionality reduction, as well as for a few other unsupervised learning applications.

Automated Machine Learning (AutoML): Google's AI project responsible for creating an AI that designs and implements its own AI, for computer vision purposes.

B

Backpropagation: a very popular training algorithm for deep learning systems and artificial neural networks in general. It involves moving the errors of the network backward, from the targets toward the inputs, and changing the weights of the various neurons based on the partial derivatives of the error function, as well as their location on the network.

Bayesian Statistics: a sub-field of Statistics that involves an alternative use of probabilities as proxies of beliefs rather than something deriving from the past. Bayesian Statistics are popular today due to their ability to work with limited data as well as their interpretability and ease of visualization.

Bias (of a model): a key characteristic of a predictive model relating to its performance. High bias means that a model is off consistently, while low bias means that it's more accurate on average. Bias is a key component of a model's performance, and it's linked to its variance in the bias-variance trade-off.

Bias-variance trade-off: a law describing the performance of a model. If a model's bias increases, its variance decreases, and vice-versa. The bias-variance trade-off plays an important role in model optimization for all predictive analytics models.

Big data: an area of computer science that is interested in the efficient processing and storage of very large amounts of data. Although defining the term changes from person to person, one can succinctly define big data as the amount of data that is big enough so that an average personal computer is unable to process.

Binarization: the data engineering process of turning a discreet variable into a binary feature.

Binning: also known as discretization, binning refers to the transformation of a continuous variable into a discreet one. Through the binning process, there is always some loss of information, something that's often acceptable as it makes for easier to handle variables.

Black box: a predictive analytics model that is not in any way transparent or comprehensible in how it arrives at its predictions. Many Machine Learning models are black boxes, contrary to Statistical models that are generally transparent.

Business Intelligence (BI): an analytics-related role that combines business-related tasks, such as researching the competition and quantifying business objectives with data analytics, usually via statistical models and visuals.

C

Centroid: the center of a cluster of data points. The coordinates of the centroids are usually among the outputs of a clustering method.

Classification: a very popular data science methodology, under the predictive analytics umbrella. Classification aims to solve the problem of assigning a label (aka class) to a data point, based on pre-existing knowledge of categorized data, available in the training set.

Classifier: a predictive analytics system geared toward classification problems.

Cloud (computing): a paradigm that enables easy, on-demand access to a network of shareable computing resources that can be configured and customized to the application at hand. The cloud is a very popular resource in large-scale data analytics and a common resource for data science applications.

Clustering: a data science methodology involving finding groups in a given dataset, usually using the distances among the data points as a similarity metric. Clustering is an unsupervised learning methodology.

Cmeans: a variant of Kmeans, using fuzzy logic for determining the cluster centers.

Collinearity: the state of a feature set whereby two or more features are strongly correlated to each other. Collinearity can be an issue with some data models, and it is usually handled with dimensionality reduction.

Computer cluster: a collection of computers sharing resources and working together, usually as a single machine. Computer clusters are very useful for tackling big data problems in-house, though more often than not, are found in data centers, forming public computer clouds.

Computer Vision: an application of artificial intelligence, where a computer can discern a variety of visual inputs and effectively "see" a lot of different real-world objects in real-time. Computer vision is an essential component of all modern robotics systems, such as self-driving cars.

Confidence: a metric that aims to reflect the probability of another metric being correct. Usually, it takes values between 0 and 1 (inclusive). Confidence is linked to Statistics, but it lends itself to heuristics and Machine Learning systems, too, since it is entirely independent as a concept. Confidence is particularly important in classification since many Machine Learning models use it for their final decision.

Confusion matrix: a k-by-k matrix depicting the hits and misses of a classifier, for a problem involving k classes. For a binary problem (involving 2 classes only), the matrix comprises various combinations of hits (trues) and misses (falses) referred to as true positives (cases of value 1 predicted as 1), true negatives (cases of value 0 predicted as 0), false positives (cases of value 0 predicted as 1), and false negatives (cases of value 1 predicted as 0). The confusion matrix is the basis of most evaluation metrics relevant to classification.

Correlation (coefficient): a metric of how closely related two continuous variables are, in a linear manner.

Cost function: a function for evaluating the amount of damage the total of all misclassifications total, based on individual costs preassigned to different kinds of errors. A cost function is a popular performance metric for complex classification problems, and it relies on the confusion matrix for its key variables.

D

Data: anything potentially containing information. Data is the prima materia of data science and data analytics in general, taking various forms. All data is eventually transformed into numbers, as it's easier to analyze it this way.

Data analytics: a general term to describe the field involving data analysis as its main component. Data analytics is more general than data science, although the two terms are often used interchangeably in the business world.

Data anonymization: the process of changing the data so that it cannot be used to identify any particular individual via the data corresponding to him or her. This involves removing or masking any personally identified information (PII) from the data analyzed.

Data engineering: the part of the data science pipeline where data is acquired, cleaned, and processed so that it is ready to be used in a data model. Most Machine Learning systems don't require a lot of data engineering as they can work on cruder forms of data all the same.

Data exploration: the part of the data science pipeline where the various variables are examined using Statistics and data visualization, to understand it better and work out the best ways to tackle it in the stages that follow.

Data frame: a data structure similar to a database table that is capable of containing different types of variables and performing advanced operations on its elements. Data frames are used in data science, particularly in combination with specialized programming libraries.

Data mining: the process of finding patterns in data, usually in an automated way. Data mining is a data exploration methodology, and it is often seen as the precursor of data science.

Data model: a data science module processing or predicting some piece of information, using existing data, after the latter has been pre-processed and made ready for this task. Data models add value and are comprised of non-trivial procedures. In AI, data models are usually sophisticated systems making use of several data-driven processes under the hood.

Data point: a single row in a dataset, corresponding to a single record of a database.

Data science: the interdisciplinary field undertaking data analytics work on all kinds of data, with a focus on big data, for mining insights or building data products. Data science includes Machine Learning, as well as other data analytics frameworks.

Data structure: a collection of data points in a structured form, used in programming as well as various parts of the data science pipeline.

Data visualization: the process of creating visuals based on the original data, or the data stemming from the data model built using the original data.

Database: an organized system for storing and retrieving data using a specialized language. The data can be structured or unstructured, corresponding to SQL and NoSQL databases. Accessing databases is a key process for acquiring data for a data science project.

Dataset: the data available to be used in a data analytics project, in the form of a table or a matrix. A dataset may need some work before it is ready to be used in a data model, though in many Machine Learning models, you can use it as-is.

Deep Learning (DL): an artificial intelligence methodology, employing large artificial neural networks, to tackle very complex problems. DL systems require a lot of data to yield a real advantage in terms of performance. A couple of powerful DL frameworks worth looking into are MXNet and Knet, both of which work with Julia.

Deterministic: a process that always yields the same result for the same outputs. For example, all descriptive statistical methods are deterministic, though most Machine Learning ones are not. A process that is not deterministic is referred to as stochastic.

Dimensionality reduction: the process of reducing the number of features in a dataset, usually through the merging of the original features in a more compact form (feature fusion), or through the discarding of the less information-rich features (feature selection). A commonly used method for dimensionality reduction is principal components analysis. Two popular Machine Learning based dimensionality reduction methods are T-SNE and UMAP.

Discretization: see *binning*.

Distribution: in Statistics, the model that describes a particular pattern in the data related to how the data is distributed across all the possible values it can take.

Distributions play a crucial role in statistical models and are plenty in number. The most well-known distribution is the Gaussian one, aka Normal Distribution.

Dropout: a method for training an ANN to mitigate over-fitting by randomly turning off some of the nodes. This enables the network to train better by introducing additional noise that helps it generalize better.

E

Ensemble: *"The process by which multiple models, such as classifiers or experts, are strategically generated and combined to solve a particular computational intelligence problem. Ensemble learning is primarily used to improve the (classification, prediction, function approximation, etc.) performance of a model, or reduce the likelihood of an unfortunate selection of a poor one."* (Dr. Robi Polikar). Ensembles may also involve Machine Learning systems, too, in order to attain better performance than a single system. Ensembles are usually black boxes.

Embedding: a low-dimensional representation of a given set of data. Embeddings are quite common in dimensionality reduction systems, particularly Machine Learning ones, such as Isomap and autoencoders. Embeddings are often referred to as meta-features.

Epoch: an iteration in the training phase of an Artificial Neural Network.

Error function: the function used for assessing the deviation of the predicted values of a Machine Learning model from the actual values (target variable). In artificial neural network models, the error function needs to be continuous.

Error rate: the complementary of accuracy rate, as a performance metric for classification systems. It denotes the proportion of data points predicted wrong. Error rate works well for balanced datasets and it's a complementary to the accuracy rate: ER = 1 – AR.

Ethics: a code of conduct for a professional. In data science, ethics involves certain practices like data security, privacy, and proper handling of the insights derived from the data analyzed.

ETL (Extract, Transform and Load): a process in all data-related pipelines, having to do with pulling data out of the source systems (usually databases) and placing it into a data warehouse or a data governance system. ETL is an important part of data acquisition, preceding any data modeling efforts. ETL is under the data engineering umbrella on a very low-level, and as a task, it is undertaken by a specialist, usually having a strong programming background.

Extreme Learning Machines (ELMs): a relatively new type of artificial neural network that is very fast to train and exhibit decent performance in predictive analytics problems. Their key characteristics are that most of the connections have random weights, apart from those of the last layer (outputs), which are optimized during the training process.

F

F1 metric: aka F1 score, a popular performance metric for classification systems, defined as the harmonic mean of precision and recall, and just like them, corresponds to a particular class. In cases of unbalanced datasets, it is more meaningful than the accuracy rate. F1 belongs to a family of similar metrics each one being a function of precision (P) and recall (R) in the form $F_\beta = (1 + \beta^2) (P * R) / (\beta^2 P + R)$, where β is a coefficient related to the importance of precision in the particular aggregation metric F_β. For the F1 metric β takes the value of 1 (i.e. precision is equally important to recall).

False Negative: in a binary classification problem, it is a data point of class 1, predicted as class 0. See *confusion matrix* for more context.

False Positive: in a binary classification problem, it is a data point of class 0, predicted as class 1. See *confusion matrix* for more context.

Feature: a processed variable capable of being used in a data science model, particularly a predictive analytics one. Features are generally the columns of a dataset.

Feature engineering: the process of creating new features, either directly from the data available, or via the processing of existing features. Feature engineering is part of data engineering, in the data science process.

Feature fusion: see *fusion*.

Feature selection: the data science process according to which the dimensionality of a dataset is reduced through the selection of the most promising features and the discarding of the less promising ones. How promising a feature is depends on how well it can help predict the target variable and is related to how information-rich it is.

Feed-forward Network: see *Multi-Layer Perceptron*.

Fitness function: an essential part of most artificial intelligence systems, particularly optimization-related ones. It depicts how close the system is getting to the desired outcome and helps it adjust its course accordingly. In most AI systems, the fitness function represents an error or some form of cost, which needs to be minimized, though in the general case, it can be anything, and depending on the problem, it may need to be maximized.

Framework: a set of tools and processes for developing, testing, and deploying a certain system. Most AI systems today are created using a framework. A framework is usually accompanied by a library/package in the programming languages it supports. In the deep learning case, for example, a framework can be a programming suite like Knet and MXNet that enables a variety of deep learning-related processes and classes to be utilized.

Functional Programming: a programming paradigm where the programming language is focused on functions rather than objects or processes, thereby eliminating the need for a global variable space. Scripts of functional languages are very modular and easy to debug. Julia is primarily a language of this paradigm.

Fusion: usually used in conjunction with the word feature (e.g. feature fusion), this relates to the merging of a set of features into a single meta-feature that encapsulates all, or at least most, of the information in these features. This is a popular method of dimensionality reduction, and it is an integral part of every deep learning system.

Fuzzy Inference System (FIS): an AI system based on Fuzzy Logic, geared toward making predictions using inference rules. An FIS is quite useful, particularly when interpretability is a concern and when opting to conserve computational resources. However, an FIS is generally limited to lower dimensionality datasets.

Fuzzy K Nearest Neighbor (FKNN): A Fuzzy Inference System developed in 1985 and improved ever since, geared toward predictive analytics applications, using a Fuzzy Logic-based approach to kNN. Modern versions of FKNN focus mostly on classification, although the same idea could be applied to regression problems, too, at least according to the original version of the ML system.

Fuzzy Logic: a term coined by Lotfi Aliasker Zadeh in 1965, referring to a different way of processing information which, unlike classical logic, also involves partial truths (instead of just the conventional black-and-white logical paradigm). Fuzzy logic uses degrees of truth as a mathematical model of vagueness. It allows for all intermediate possibilities between digital values of YES and NO, much like how a human will assess the nature of a situation in full color and multi-polar fashion, rather than a bi-polar, monochrome way. Fuzzy logic is a well-established aspect of artificial intelligence.

G

GDPR: an EU legislation related to the acquisition of private data (mostly PII) and the need for permissions from the customers regarding the use of that data. Even though GDPR applies to EU countries only, several other countries have adopted its policies, partly because you still need to abide by them if you want to do business with people living in the EU.

Gen: a novel programming language stemming from Julia, used for probabilistic programming in various fields, particularly AI Gen exists as a Julia package. It aims to create an abstraction of all the mathematical aspects of the probabilistic models used in data analytics. Gen was developed by a team at MIT.

Generalization: a key characteristic of a data science model, where the system is able to handle data beyond its training set in a reliable way. A proxy to good generalization is having similar performance between the training set and a testing set, as well as consistency among different training-testing set partitions of the whole dataset.

Generative Adversarial Networks (GANs): a kind of minimalist ensemble of ANNs that is used for generating new data, particularly related to multimedia and text. GANs are comprised of two ANNs, one geared toward classifying something and the other toward fooling the first one by creating data that cannot be classified correctly. As the two ANNs train (simultaneously), they both become better, yielding somewhat realistic new data that can be used to enhance the original data.

GPU (Graphics Processing Unit): a specialized component of a computer's hardware, designed for processing data related to the computer's display such as image-like data. GPUs can be leveraged to obtain additional computing power for resource-demanding tasks, such as artificial intelligence systems, as in the case of Deep Learning.

Graph: A kind of dimensionless structure that is an abstraction of the objects involved in a process as well as their relationships (connections). It is characterized by nodes and arcs, representing the objects and the connections, respectively. The latter also carries other characteristics too, such as weights, corresponding to the strength of each connection.

Graph analytics: a data science methodology making use of Graph Theory to tackle problems through the analysis of the relationships among the entities involved.

H

Hadoop: a distributed data storage technology developed by Apache Foundation. Hadoop is geared toward big data and comprises a variety of data analytics tools.

Heuristic: an empirical metric or function that aims to provide some useful tool or insight, to facilitate a method or project of data science or artificial intelligence.

Heuristics are entirely data-driven and focus on performing a very specific task in an efficient and scalable manner.

I

IDE (Integrated Development Environment): a system designed for facilitating the creation and running of scripts as well as their debugging. Jupyter is a popular IDE for data science applications.

Insight: a non-obvious and useful piece of information deriving from the use of a data science model on some data.

Interpretability: the ability to more thoroughly understand a data model's outputs and derive how they relate to its inputs (features). Lack of interpretability is an issue for deep learning systems as well as many Machine Learning systems in general. Interpretability is often referred to as transparency.

J

JLD: a file type used in Julia for storing dataset, usually in the form of a dictionary data structure. The latest version of this type, JLD2, is the one recommended due to its speed and ease-of-use. To access JLD2 files you need both of the packages JLD2 and FileIO loaded into memory.

Julia: a modern programming language of the functional programming paradigm, comprising characteristics for both high-level and low-level languages. Its ease of use, high speed, scalability, and sufficient amount of packages, making it a robust language well-suited for data science. After v. 1.0 of the language was released, it has been officially production-ready in a wide variety of organizations.

Jupyter: a popular browser-based IDE for various data science languages, such as Python and Julia. Jupyter's most popular variant is Jupyter Notebook, although there is another variant developed fairly recently called JupyterLab.

K

K-fold Cross Validation: a fundamental data science experiment technique for building a model and ensuring that it has a reliable generalization potential. K-fold cross-validation is used in combination with a performance metric, such as mean square error.

K-means: a popular clustering method based on distances among the data points involved. Its key parameter, K, corresponds to the number of clusters we wish to have in the clustering process.

Knet: a deep learning framework, developed entirely in Julia.

KNN: a basic Machine Learning model, relying on distances. KNN is one of the most widely used transductive models, although it is not as relevant today due to its various limitations. FKNN is a popular variant of KNN, employing fuzzy logic.

L

Labels: a set of values corresponding to the points of a dataset, providing information about the dataset's structure. The latter takes the form of classes, often linked to classification applications. The variable containing the labels is usually used as the target variable of the dataset.

Layer: a set of neurons in an artificial neural network. Inner layers are usually referred to as hidden layers and consist mainly of meta-features created by the system.

Library: see *package.*

Logistic function: see *sigmoid function.*

M

Machine Learning (ML): a set of algorithms and programs that aim to process data without relying on statistical methods. ML is generally faster, and some methods of it are significantly more accurate than the corresponding statistical ones, while the assumptions they make about the data are fewer. There is a noticeable overlap between ML and artificial intelligence systems designed for data science.

Manifold: a mathematical technique that provides a view of a flat feature space, having a similar behavior to that of a Euclidean space. Manifolds are used in a variety of dimensionality reduction methods, such as Isomap and UMAP.

Mapping: the process of connecting a variable or a set of variables, to a variable we are trying to predict (aka target variable). Mappings can be analytical using a mathematical function, or not, such as employing a set of rules, or a network of functions, as in the case of an artificial neural network. Mappings are inherent in every data model.

Mean Square Error (MSE): a popular performance metric used for regression problems. It involves taking the difference between the target variable and the predicted values of the target variable, squaring it, and then taking the average. The model having the smallest such error is usually considered the better one.

Meta-features (aka super features or synthetic features): high-quality features that encapsulate larger amounts of information, usually represented in a series of conventional features. Meta-features are either synthesized in an artificial intelligence system or created through dimensionality reduction. Generally, a meta-features has a stronger predictive potential than each one of the features it derives from.

Metadata: data about a piece of data. Examples of metadata are timestamps, geo-location data, data about the data's creator, and notes.

Methodology: a set of methods and the theory behind those methods, for solving a particular kind of problem in a certain field. Methodologies of data science include classification and regression, while for AI we have methodologies like deep learning and autoencoders.

Model: see *data model*.

Model Maintenance: the process of updating or even upgrading a data model, as new data becomes available or as the assumptions of the problem change.

Multiple dispatch: a feature of some programming language, enabling the use of the same command to handle different tasks, based on the inputs provided. This greatly mitigates the overhead of function names, while it makes the use of the language smoother and more user-friendly.

Multi-Level Perceptron (MLP): a deep learning system that comprises a series of layers of neurons, much like a normal ANN, but larger. It is often referred to as a feed-forward network, and it's the first system in the deep learning family to have been developed. MLPs are great for various standard data science problems, such as classification and regression.

MXNet: a deep learning framework developed by Apache. MXNet is linked to Amazon, although it can run on any cloud computing service. Its main API is called Gluon, and it's part of the main package of MXNet. There are several such packages in different programming languages, each one an APIs for that language. MXNet can support more programming languages than any other deep learning framework.

N

Natural Language Processing (NLP): a text analytics methodology focusing on categorizing the various parts of speech for a more in-depth analysis of the text involved.

Narrow AI: see *weak AI*.

Network: a collection of nodes, forming a data structure often used in artificial intelligence systems. Computer Science networks are known as graphs in Mathematics.

Neuron: a fundamental component of an artificial neural network, usually representing an input (feature), a meta-feature, or an output. Neurons in a network-based AI system are organized in layers.

Node: a key part of a graph, namely a connector of various arcs. In Machine Learning, nodes are found in decision trees (particularly where the splits are done) and ANNs. In the latter case, they are referred to as neurons.

Noise: all the parts of the dataset that don't add any value to data science work due to their random nature. Noise is usually handled in the data engineering phase and is in contrast with the signal of the dataset.

Non-Negative Matrix Factorization (NMF or NNMF): an algebraic technique for splitting a matrix containing only positive values and zeros, into a couple of matrices that correspond to meaningful data, useful for recommender systems.

Normalization: the process of transforming a variable so that it is of the same range as the other variables in a dataset. This is done through statistical methods primarily and is part of the data engineering stage of the data science pipeline.

O

Object-Oriented Programming (OOP): a programming paradigm where all structures, be it data or code, are handled as objects. In the case of data, objects can have various fields (referred to as attributes), while when it comes to code, objects can have various procedures (referred to as methods).

Objective function: see *fitness function*.

Optimization: an artificial intelligence process, aimed at finding the best value of a function (usually referred to as the fitness function) given a set of restrictions. Optimization is key in all modern data science systems, particularly Machine Learning ones. Although there are deterministic optimization algorithms, most of the modern algorithms are stochastic.

Optimizer: a (usually AI-based) system designed to perform optimization.

Over-fitting: the case whereby a model is too specialized to a particular dataset, yielding excessive variance. Its main characteristic is good performance for the training set and poor performance for any other dataset. Over-fitting is a characteristic of an overly complex model.

P

Package: a set of programs designed for a specific set of related tasks, sharing the same data structures, and freely available to the users of a given programming language. Packages may require other packages to function, which are called dependencies. Once installed, the package can be imported in the programming language and used in scripts.

Paradigm: an established way of doing things as well as the set of similar methodologies in a particular field. Paradigms change very slowly, but when they do, they are accompanied by a change of mindset and frequently new scientific theory.

Parallelization: the case of using parallel computing for performing tasks simultaneously. Parallelization is accomplished using CPUs or GPUs as nodes performing certain tasks (aka workers) and then combining the results using a master node. Modern data science would not have been possible without parallelization, though the latter has applications in other fields, including artificial intelligence.

Pdf (probability density function): a function the can yield the probability of a given range of values for a particular distribution, particularly useful in Statistics. Every distribution has a unique pdf while taking the natural logarithm of that function yields the logpdf of that distribution.

Perceptron: a rudimentary artificial neural network, comprising a single neuron. When it comes to classification, a single perceptron can only handle very simple problems as it fails to generalize non-linear class boundaries.

Performance metric: a heuristic geared toward evaluating the performance of a data model. Performance metrics are an essential part of validating a model, making sure it is ready for being put into production. Also, every methodology has its own performance metrics. Popular performance metrics for Classification are Accuracy and F1 score, while for regression the Mean Square Error is usually used.

Personally Identifiable Information (PII): information that can be used to pinpoint a particular individual, thereby violating his/her privacy. PII is an important ethical concern in data science and may not be so easy to tackle since it often relies on a combination of variables.

Pipeline: also known as workflow, it is a conceptual process involving a variety of steps, each one of which can comprise of several other processes. A pipeline is essential for organizing the tasks needed to perform any complex procedure (often non-linear) and is very applicable in data science (this application is known as the data science pipeline).

Plot: a visual, representing data graphically. Plots are essential for data exploration, among other things.

Population: the whole of the data (usually the data available at us). The whole population is rarely used, due to the computational cost this entails, which is why a sample (or a series of samples) is used instead.

Possibilistic: related to memberships (possibilities) for modeling uncertainty. Possibilistic methods have existed since the 1970s, with the advent of Fuzzy Logic as a framework for dealing with uncertainty in a methodical and mathematically rigorous manner that has no dependency whatsoever to statistical modeling, which are probabilistic in nature.

Precision: a performance metric for classification systems, focusing on a particular class. It is defined as the ratio of the true positives of that class over the total number of predictions related to that class. Precision is complementary to recall as a performance metric for classification.

Predictive analytics: a set of methodologies of data analytics, related to the prediction of certain variables. It includes a variety of techniques such as classification, regression, time-series analysis, and more. Predictive analytics are a key part of data science and one that adds the most value in data science projects.

Principal Components Analysis (PCA): a statistical method for dimensionality reduction through the fusion of features to form a series of meta-features known as principal components (PCs). Information content in PCA is measured through the variance statistic, while the method is good at restoring the original feature set through a subset of the PCs, with minimal distortion. PCA is often used in data science as a dimensionality reduction technique, and it is deterministic in nature.

Probabilistic: related to probabilities for modeling uncertainty.

Pruning: the process of cleaning up code so that unwanted solutions can be eliminated. However, with this process, the number of decisions that can be made by machines is restricted.

Python: a widely used object-oriented programming language, typically used for data science, as well as artificial intelligence applications geared toward data analytics.

Q

Quantum computing: a new computing paradigm that leverages quantum effects and the use of qubits for carrying out computing tasks potentially faster. Not all problems can be solved efficiently with quantum computing, but it is believed that Machine Learning, particularly models based on AI methods, has a lot to benefit from quantum computing.

Qubits: short for quantum bits, a new kind of information unit, in the quantum computing paradigm. Unlike conventional bits, which can be either 0 or 1 only, qubits can take 0, 1, and both at the same time, leveraging the superposition concept from Quantum Physics.

R

Recall: a performance metric for classification systems, focusing on a particular class. It is defined as the ratio of the true positives of that class over the total number of data points related to that class. Recall is complementary to precision as a performance metric for classification.

Recommender system (RS): also known as a recommendation engine, a RS is a data science system designed to provide a set of similar entities as the ones described in a given dataset, based on the known values of the features of these entities. Each entity is represented as a data point in the RS dataset.

Rectified Linear Unit (ReLU) function: a commonly used transfer function for an artificial neural network. ReLU is often preferred due to its low computational cost and high performance in terms of accuracy. It is defined as $f(x) = max(w^*x+b, 0)$ and takes values between 0 and infinity.

Regression: a very popular data science methodology, under the predictive analytics umbrella. Regression aims to solve the problem of predicting the values of a continuous variable corresponding to a set of inputs, based on pre-existing knowledge of similar data, available in the training set.

Regressor: a predictive analytics system geared toward regression problems.

Reinforcement Learning (RL): a Machine Learning methodology that involves the use of feedback on the model, as a means to learn interactively from its environment. RL is particularly useful in robotics applications as well as games, due to the minimal supervision that is required during the learning task. However, a properly defined objective function needs to be in place as well as exact values of the reward or punishment involved for the various actions taken (which also need to be defined beforehand). RL is almost always applied through the use of AI-based models.

Robotic process automation (RPA): a method or technique enabling the integration of software or the automation of work processes, through the use of the same user interface. RPA is often used in combination with somewhat "clever" scripts, but it doesn't qualify as AI, at least for the majority of use cases. Also, RPA has nothing to do with Machine Learning.

Robust (model): a state whereby a model is not over-fit, yielding a consistent performance across different data samples. Robust models are considered reliable enough to be used in production.

ROC analysis: short for Receiver Operating Characteristics analysis, an evaluation method for binary classifiers, examining how the False Positive Rate and the True Positive Rate (Recall) relate with each other. The result of this analysis usually takes the form of a curve (aka ROC curve), while the area under that curve can be used as a holistic evaluation metric (aka AUC). Yet, even without the AUC metric, the ROC analysis is useful as it examines how the different values for the decision threshold (often depicted as λ) affect the outcome of the classifier, helping us decide what the best trade-off between FP and FN is, for the problem at hand.

ROC curve: a curve representing the trade-off between true positives and false positives for a binary classification problem, useful for evaluating the classifier used. The ROC curve is usually a zig-zag line, depicting the true positive rate for each false positive rate value. The area under the curve is also used as an evaluation metric, namely AUC.

S

Sample: a limited portion of the data available, useful for building a model, and (ideally) representative of the population where it belongs.

Sampling: the process of acquiring a sample of a population using a specialized technique. Sampling is very important to be done properly, to ensure that the resulting sample is representative of the population studied. Sampling needs to be unbiased—something usually accomplished by making it random.

Scala: a functional programming language, very similar to Java, that is used in data science. The big data framework Spark is based on Scala.

Semi-supervised learning: a fairly new Machine Learning methodology that involves the use of both labeled and unlabeled data in a predictive analytics setting. Semi-supervised learning helps eliminate biases in the data and drive down the costs of a data science project.

Sensitivity Analysis: the process of establishing how stable a result is or how prone a model's performance is to change if the initial data is different. It involves several methods, such as resampling and "what if" questions. For binary classification problems, it involves additional tools such as ROC analysis.

Sentiment analysis: a natural language processing method involving the classification of a text into a predefined sentiment, or the figuring out of a numeric value that represents the sentiment polarity (how positive or how negative the overall sentiment is).

Sigmoid function: a mathematical function of the form $f(x) = 1 / (1 + \exp(-(w^*x + b)))$. Sigmoids are used in various artificial neural networks, such as Deep Learning networks, as transfer functions. Sigmoid takes values between 0 and 1, not inclusive. This is sometimes referred to as the logistic function as it features in logistic regression.

Signal: the underlying gist of a dataset, which depicts the information within the data at hand. The signal is not easily measurable, and it's often contradicted to the noise of the data.

Softmax function: a transfer function sometimes used in a deep learning network. It is a simpler version of the sigmoid function, where the bias parameter (b) is missing from the equation. Softmax takes values between 0 and 1, not inclusive.

Spark: a powerful big data platform, built on top of Hadoop, using Scala. Spark features high efficiency and has its own data structures as well as Machine Learning libraries.

Stochastic: something that is probabilistic in nature (i.e. not deterministic). Stochastic processes are common in most artificial intelligence system and other advanced Machine Learning systems.

Strong AI: an area of AI development that is working toward the goal of making AI systems that are as useful and skilled as the human mind.

Supervised learning: a set of data science methodologies where there is a target variable that needs to be predicted. The main parts of supervised learning are classification, regression, and reinforcement learning.

T

Tanh function: the hyperbolic tangent function. It is of the same family as the sigmoid, and it is sometimes used as a transfer function for deep learning networks. It is defined as $f(x) = (\exp(x) - \exp(-x)) / (\exp(x) + \exp(-x))$ and takes values between -1 and 1, not inclusive.

Target variable: the variable of a dataset that is the target of a predictive analytics system, such as a classification or a regression system.

TensorFlow: a deep learning and high-performance numerical computation library. Initiated by Google and improved by a very large open source community, TensorFlow is by far the most popular deep learning framework today, even if there are other, better DL frameworks.

Time-series Analysis: a data science methodology aiming to tackle dynamic data problems, where the values of a target variable change over time. In time-series analysis, the target variable is also used as an input in the model.

TPU: short for TensorFlow Processing Unit, a TPU is a hardware computer component geared toward making TensorFlow operations faster and more efficient.

Testing set: the part of the dataset that is used for testing a predictive analytics model after it has been trained and before it is deployed. The testing set usually corresponds to a small portion of the original dataset.

Training algorithm: the algorithm used for training a deep learning system (or a predictive analytics model in general). It entails figuring out which nodes to keep and what weights their connections have, so as to obtain a good generalization for the problem at hand. Back-propagation is an established training algorithm, suitable for various kinds of artificial neural networks, including deep learning systems.

Training set: the part of the dataset that is used for training a predictive analytics model before it is tested and deployed. The training set usually corresponds to the largest portion of the original dataset.

Transductive model: a predictive analytics model relying on a distance or similarity metric for its predictions. A typical, albeit fairly obsolete transductive model, is kNN.

Transfer function: component of an artificial neural network, corresponding to the function applied on the output of a neuron before it is transmitted to the next layer. A typical example is the sigmoid function, though ReLU is often used in practice too. Transfer functions are sometimes referred to as activation functions.

Transfer learning: the process of using a model trained on completely different data to tackle a different problem. Transfer learning is made possible due to the great generalization of that model, and it's particularly popular in AI-based models.

Transparency (in a Machine Learning model): an important characteristic of a model, related to its ability to explain how it arrives at a particular conclusion as well as how confident it is about its predictions. Lack of transparency is often referred to as being a "black box." Transparency is often referred to as interpretability.

True Negative: in a binary classification problem, it is a data point of class 0, predicted as such. See *confusion matrix* for more context.

True Positive: in a binary classification problem, it is a data point of class 1, predicted as such. See *confusion matrix* for more context.

U

Under-fitting: the state of a model whereby it has insufficient variance, leading to substandard performance, consistently. Under-fitting is a sign of an overly simplistic model characterized by high bias.

Unit testing: a level of software testing where the various individual components of a program are tested to ensure that they operate as designed. Unit testing is sometimes referred to as component testing, and it can be useful in Machine Learning projects too.

Unsupervised learning: a set of data science methodologies where there is no target variable that needs to be predicted. The main unsupervised learning methods are clustering and dimensionality reduction.

V

Variable: a column in a dataset, be it in a matrix or a data frame. Variables are usually turned into features after some data engineering is performed on them.

Variance (of a model): a key characteristic of a predictive model relating to its performance. A model of high variance is bound to be unstable and, as a result, unreliable. Low variance is generally better, though the model's overall performance must take into account its bias too.

Variance (of a variable): a metric related to how the values of the variable vary in relation to its average. Variance is important in most statistical models, as well as a metric for describing a variable.

W

Weak AI: also known as *narrow AI*, weak AI refers to a non-sentient computer system that operates within a predetermined range of skills and usually focuses on a singular task or small set of tasks. All AI in use today is weak AI.

Workflow: see *pipeline*.

Wrapper: a method or framework that acts as a proxy for blending various other methods together. Wrappers are particularly useful in complex problems involving lots of functions and processes.

Answers

Chapter 1

1. What makes Julia a functional language?

A: The fact that it makes use of a function-based approach to programming, with a focus on keeping the variables involved in each function independent, unless stated otherwise (i.e. have a "global" scope). Also, Julia manages its memory automatically, and although there was a garbage collection function in previous versions of it, this feature doesn't exist anymore.

2. Why is Julia relevant as a data science language?

A: Because it is high level, fast, and has many packages that are designed for data science work. Being high level makes it easy to use and prototype in, while its high speed enables it to be useful when tackling large amounts of data (this combination of traits also solves the two-language problem). Finally, the fact that it has many data science packages makes tackling data science tasks efficient and straightforward, even if you are not an adept programmer.

3. Do you need to be a programmer to learn and master the Julia language?

A: No. Julia is a straightforward high-level language, so learning it isn't that challenging, no more than learning Python or R anyway. As for mastering it, it does take some time, but you don't need to have special training in computer science to do this. If you do, however, it can speed up the process considerably.

4. Isn't Julia too new to be reliable as a programming language?

A: No, although before the summer of 2018, it was fairly new still. However, after its v. 1.0 release, it is production-ready, plus the numerous packages it has make it a useful tool for many applications. As for data science work, it's been reliable since v. 0.3 or so.

5. How does Julia compare with Java in terms of performance?

A: It is slightly faster, based on a series of benchmark tests. Also, its performance across these tests is more consistent than that of Java. Note that these benchmark tests were conducted using Julia 1.0.0 and Java 1.8.0_17. You can learn more about them at the corresponding web page: https://bit.ly/3boqLaO.

Chapter 2

1. When would you use an IDE like Atom in your work?

A: If it is required to write code that is to be used in various projects, for example, a function for processing data or some ETL process. Also, if the script is going to be elaborate, Atom would be a good choice due to the functionality the IDE offers.

2. Can you use a basic text editor for creating or editing your Julia scripts?

A: Yes, although this may not be the most efficient way of carrying out this task. However, if no other option is available, a basic text editor would do just fine, especially for simpler scripts.

3. What kind of file would you use to store the code of a multi-purpose function you have created, to use in your data science projects?

A: A .jl file, so that it can be used in different programs and projects.

4. Can you use Jupyter offline?

A: Yes. In fact, it's rarely used online, unless you need to make use of JuliaBox.

5. Is Jupyter better than other IDEs?

A: No, just different. For certain tasks, it's ideal, but other IDEs have advantages of Jupyter, such as being more lightweight and handling scripts from many programming languages, while Jupyter has a more limited repertoire.

6. Can you run Julia on your mobile device?

A: Yes, though it's not particularly common. In that case, it's better to do so through the use of JuliaBox.

Chapter 3

1. What's the point of using pre-made packages in Julia if we can code everything from scratch without any serious compromise in performance?

A: Because not everyone who knows Julia can code a reliable script for a specific task, while ensuring it works for a variety of use cases. Also, coding everything from scratch can be time-consuming. At the very least, the plotting packages ·should be used.

2. Can you trust 3rd party packages for data science applications in Julia?

A: It depends on their source. However, it is best to rely on packages in the official Julia repository, unless you know the creator of the 3rd party package and trust their code.

3. What's the best data science-related package in Julia?

A: The one that helps you best with your task at hand. The stars system on GitHub may provide an indication of a package's popularity, but at the end of the day, it's really up to you to decide which package is best suited for the task you need to do.

4. What would you do if none of the packages presented in this chapter are suitable for your data science-related project?

A: Research the package library on your own and try to find an alternative package to use, or if nothing is suitable enough, code something from scratch.

5. Do you need to know the algorithms behind the programs in the data science related packages to use them?

A: No, but it's good if you do. Otherwise, you may not be able to fine-tune the methods properly or interpret the results. Knowing the algorithms behind the programs can also help with troubleshooting any issues that arise.

6. What can you do if you come across an issue (bug) in one of the official packages of the Julia ecosystem?

A: Report it to the Julia community so that someone will fix it eventually. Alternatively, attempt to fix it and share the fixed version with everyone through Github.

7. Do you need all of the packages presented in this chapter for your data science work?

A: No, but it's always good to be aware of them since they may come in handy in a data science project in the future.

Chapter 4

1. Why is Machine Learning important these days?

A: Because despite what the Statistics fans claim, statistical models don't cut it for many modern real-world applications, particularly related to predictive analytics. Machine Learning models, although relatively opaque in general, tend to have better performance and are as fast to run.

2. What's the key difference between kNN and a statistical model?

A: kNN makes use of distances between pairs of data points to assess similarity, to gauge the labels of the data points it needs to predict (test set), based on the data points it already knows (training set). A statistical model creates a linear mapping between the inputs (features) and the outputs (target variable) and optimizes the corresponding parameters, all while making a few assumptions about the data along the way. kNN uses the features in tandem (through the aforementioned distances) and makes no assumptions whatsoever about the data. Also, statistical models are interpretable while kNN is not.

3. How does Machine Learning relate to AI?

A: There is an overlap between the two. Mainly ANN-related models, which although AI in nature, when used in data science, do so through Machine Learning. Also, fuzzy logic systems, although technically under the AI umbrella, are usually applied in a Machine Learning context.

4. Is there a discrepancy between what we as data scientists think of about Machine Learning and what a businessperson may think?

A: Yes. We as data scientists tend to view things at a more granular level when it comes to technology. A businessperson, on the other hand, is bound to have a birds-eye view, conflating Machine Learning with AI as well as other data science-related topics. Depending on their understanding of the subject, they may view Machine Learning as a data science methodology, an AI sub-field, or even some form of wizardry! From their standpoint, it's all about yielding useful results, so what Machine Learning is exactly is not as important.

5. Is it better to use a Machine Learning method instead of a model-based one? Explain.

A: Sometimes, depending on the requirements of the project. If performance (e.g. accuracy rate or some other evaluation criterion) is of the utmost importance, Machine Learning would be a better choice for the method to use. However, if interpretability is required, a model-based approach would be preferable.

6. Does the Machine Learning know-how from a decade ago apply to today? Why?

A: To some extent, yes. However, on the AI side, things have evolved greatly since the previous decade, so any ANN-related know-how may be stale or even obsolete. For the basic Machine Learning models, however, things haven't changed much, if at all.

Chapter 5

1. How can you make sure that a given package X is up-to-date so that you can use it for your Machine Learning task?

A: Run the *update* command at the Pkg prompt and wait for a bit (how much depend on how many packages are installed and how many of them need updating):

(v1.x) pkg> update

2. What's the difference between a data frame and a matrix?

A: A data frame is a more modern data structure that is capable of containing data of different types (e.g. have a column comprised of integers, another one comprised of strings, and another column comprised of floats). Also, the data frame has numbered rows and metadata related to the variables in it (i.e. their names, aka a header row). The matrix data structure is homogeneous and generally more versatile, particularly when it comes to mathematical operators as matrices tend to be numeric.

3. How would you use the CSV package to handle a data file that contains headers? What happens to the header metadata?

A: By making use of the header parameter when loading the data:

```
MyDataFrame = CSV.read("datasource.csv", header = true);
```

The header metadata will be part of the data frame accessible through the names command. Note that all data stored there will be in the Symbol data type, which is different from normal strings.

4. Why would you want to normalize the data at hand? Are there other ways to do so apart from the ones mentioned in this chapter?

A: Because many predictive analytics methods require the features to be of the same scale. Using the features in their original scales is bound to confuse these systems, leading to inaccurate predictions. There are various ways to perform normalization beyond the two main ones that are mentioned in the chapter.

5. Can you combine clustering and classification for better results? What about clustering and regression?

A: Yes, although it may not necessarily help the result significantly. The idea is to use the clustering labels as an additional feature or set of features. This can be applied in both classification and regression.

6. What's the most important Machine Learning utility introduced in this chapter?

A: All Machine Learning utilities are important, and although some are more important in certain scenarios, there is no clear "winner" as to which one is the most important.

7. Are there any assumptions you need to make about the data in order to use the methods introduced in this chapter?

A: Not really. Naturally, you would want to make sure that the packages used are up-to-date, but other than that, the data can be used without having to go through an assumptions checklist as when working with statistical methods.

8. Can you work a data science project using solely Machine Learning packages? Explain.

A: Yes, although this will compromise the quality of your work. It's best to use some statistical packages too for certain tasks, such as normalization and descriptive Statistics, at least for the data exploration stage.

Chapter 6

1. Does it matter which package you use for a particular Machine Learning model, in Julia?

A: Yes, since some packages are less efficient than others because of their learning curve and code complications. That's why it's best to know several packages and use the ones that you are most comfortable with or are most relevant to the problem at hand.

2. Could you use random sampling over and over again instead of K-fold cross-validation, when evaluating a predictive analytics model?

A: Yes, but you would have to have several samples in order to reliably gauge the performance of the model tested. K-fold cross-validation is just more efficient, though even that may not be 100% reliable in some cases.

3. How can the output of a clustering algorithm be evaluated?

A: Usually a heuristic like Silhouette Width (aka Silhouette score or just Silhouette) is used. The higher its value, the better the clustering result. However, it's not a bad idea to use K-fold cross-validation too.

4. What's the most important thing in a hands-on Machine Learning project?

A: The narrative / story of the whole project, binding its different components together into a coherent whole. The comments and other supporting text in the code are both relevant to this. However, different managers may put emphasis on different aspects of the ML project, so it's best not to rely 100% on this rule of thumb.

5. What other questions could you ask for this particular data science project?

A: A series of additional questions could be asked. Some of these are:

- Is there a difference in the quality level of red and white wines?
- What's the most important predictor of the quality of a wine?
- What are the key differences (in terms of features) between red and white wines?
- What are the most important groups of wines, for red wines?

Chapter 7

1. How would you parametrize the model showcased in this chapter even further?

A: Explore the option of having a parameter for the distance metric used, the power the distances are raised to, as well as the strength each feature would have in the whole matter, through a forced re-scaling of these features.

2. What do you need to do regularly in order to maintain the usefulness of a custom-made model like this one?

A: Make sure that there is always relevant data around, eliminate any redundant data points as new data gathers, and ensure that the model's performance is consistent as the training set is re-calibrated.

3. Could you use pre-made code (e.g. from a different code-base) in your custom-made model, to save time? What considerations do you need to have when doing so?

A: Yes, from a relevant package in Julia, for example. However, lots of changes would need to take place, so this option would be best limited to models that are too complex to code from scratch.

4. How would you prove to someone that this custom-made model is any good, for the task it is designed for?

A: Perform a series of unit tests as well as performance tests using a variety of datasets. Also, compare the performance results to those of a baseline model of the same family of models.

5. Can a custom-made model be comprised of other models (e.g. in an ensemble setting)?

A: Yes. In fact, this is more common than people think. Even the vanilla flavor ANNs (MLPs) are basically a group of perceptron models linked together in an ensemble setting, even if they are not referred to as such.

6. When would you use a programming library, and what considerations would you have when doing so?

A: When that library is shown to be well-maintained, optimized for performance, and contains essential functions for the project at hand. Considering whether this library has a large number of people involved in improving it and how long it's been in the repository is very important. Also, the documentation plays a significant role in all this since poorly documented libraries are bound to stay in the fringes of the Julia ecosystem.

7. Based on what has been discussed in this chapter, would a report resembling the paper used here be effective? How would you supplement it to make it more comprehensive and more comprehensible?

A: No, since it's too vague, too high-level, and lacks essential programming considerations, such as algorithm complexity, resources required, and proper evaluation using more than one performance metric and more than one alternative algorithm to compare. Apart from addressing these obvious issues, it would better to add a couple of examples (particularly with graphics explaining them), showcasing how FKNN is different than the original KNN algorithm, what limitations it has, and how to best implement the algorithm.

8. When would you avoid relying on an academic paper showcasing a novel algorithm, relevant to Machine Learning? Why?

A: When the paper doesn't contain any pseudo-code or when the arguments made toward the values of the Machine Learning system are unconvincing. Relying on a paper that is too vague (as it's often the case with many conference papers, for example) can lead to an inaccurate and sub-optimal implementation of the model, as well as lots of wasted time and computational resources. That's why it's best to do some research on the available research resources before you start implementing anything. The same algorithm that may seem very challenging to implement based on a particular paper may actually be quite easy based on another paper.

9. Could you use ROC analysis in the case of a multi-class classification problem?

A: Yes, if we are interested in a particular class only and deal all the other classes as a single class. This can be done through one-hot-encoding for that class of interest, effectively turning the problem into a binary classification one. Afterward, we can apply ROC analysis as usual.

Chapter 8

1. What's the advantage of UMAP, LLE and Isomap over PCA?

A: All three of these methods capture the non-linear relationships of the dataset, something that PCA doesn't do as it's a linear method. Also, when reducing the dimensionality to just two dimensions, the reduced dataset is still useful for visualization and even for some basic models like clustering.

2. Would you ever use a dimensionality reduction method on a dataset with fairly few variables? Explain.

A: Yes, if the variables are highly correlated with each other. Most likely, the dimensionality won't be reduced much. However, the whole process may still improve the dataset, especially if we plan to use the reduced dataset in certain models that require the features to be independent of each other.

3. What's the main disadvantage of UMAP, LLE and Isomap in relation to PCA and ICA?

A: All three of them are stochastic methods, so every time we run them they yield different results. PCA and ICA, on the other hand, are deterministic, so the results are always the same. This makes replicating the dimensionality reduction process much easier, in many cases.

4. Are there any AI-based dimensionality reduction methods? If so, which ones?

A: Yes, autoencoders, for example. Such methods help in optimizing a set of metafeatures so that they can reproduce the original features, even though they are fewer in number. Also, these metafeatures are non-linear combinations of the original features, something that often ensures a more efficient reduction.

5. How do UMAP, LLE and Isomap compare to feature selection methods based on the target variable?

A: They are not as good, since the latter makes use of additional information, found in the target variable. However, you can combine feature selection and UMAP/LLE/Isomap (in that order) for optimal results.

6. What's UMAP's main advantage over Isomap?

A: UMAP is significantly faster than Isomap and many other distance-based dimensionality reduction techniques.

7. Could different dimensionality reduction methods be used in tandem? What would be a good strategy for that?

A: Yes, particularly in cases where the dimensionality is too large to be handled efficiently with a single dimensionality reduction method. A good way to do that would be to apply a fairly fast method to get rid of the obviously redundant features (e.g. feature selection or some filtering technique) and then use a more powerful dimensionality reduction method on the remaining features (e.g. UMAP or PCA). If you want to use a manifold-based method in combination with PCA, it would be best to use PCA first as it's faster and then the manifold-based method on the meta-features PCA yields.

Chapter 9

1. How can you improve the performance of a model in terms of bias?

A: Add more features to your dataset, particularly features stemming from different sources, unrelated to the existing ones, to ensure higher diversity in your training data. Also, opt for a more sophisticated model (e.g. a non-linear one or an ensemble).

2. How can you improve the performance of a model in terms of variance?

A: Remove excessive features or combine features (e.g. through a process like PCA or UMAP). Also, opt for a simpler model (e.g. a linear one). If using an ensemble, exploring cases with fewer components or different parameters in case over-fitting takes place in all these components in the first place would make sense.

3. How can you improve the performance of a model in terms of both bias and variance?

A: Pick a better model, such as an AI-based one or an ensemble. Also, refine the dataset through additional data engineering, particularly when it comes to the features used.

4. Some people claim that ensemble models are better because they don't over-fit. Is there any truth in that statement?

A: There is some truth to that in the sense that they don't over-fit as easily as other models. However, they still run the risk of over-fitting if their parameters are not chosen carefully or if the data used for training them is too complex (e.g.

having an excessive amount of variables). Some data engineering can help mitigate the risk of over-fitting.

5. What's better to have in a predictive model, higher bias or higher variance? Explain.

A: There is no rule-of-thumb here, since both options are undesirable. However, if you err toward higher bias that's less of a show-stopper since the model may still yield some value, while a high variance model is bound to be too unstable to trust. Nevertheless, a high variance model is closer to the "sweet spot" of bias-variance, so if you tweak it a little, it is bound to improve faster than a high bias model.

6. How can a heuristic improve a Machine Learning model used for predictive analytics?

A: It can do so in various ways, such as through increased quality of training data (e.g. through better data engineering and assessment of features used), better decisions regarding the predictions (again through better assessment of the data while it's processed by the model), and through better evaluation for the model (e.g. through more accurate analysis of a confusion matrix).

7. Are heuristics found in Machine Learning only? Give a couple of examples or counter-examples.

A: Yes and no. Yes, in terms of the name "heuristic" which is associated with Machine Learning and AI. No, in terms of the essence of a heuristic, since Stats has a bunch of ad-hoc metrics often referred to as "Statistics" to facilitate processes and methods. In the well-known Chi-square test, for example, the chi-square statistic is a heuristic that's analyzed thoroughly and utilized to make inferences about the independence of the variables analyzed in that test. The

Pearson's Correlation metric is also a Stats heuristic, and a very popular one at that.

8. How are AI-related models related to Statistics? Explain.

A: Generally, they are not. However, many attempts have been made to link the two fields in a way that makes the former more scientifically valid, at least in the eyes of those who have deified Statistics for some reason. Nevertheless, there are some AI models that borrow a lot from Statistics in the sense that they make use of distributions and try to figure out the best parameters of these distributions to model the data better (e.g. Variational Autoencoders).

9. What's the best way to build an ensemble to have better performance overall?

A: Get component models that have mutually different errors in their predictions—ideally, these models would be inherently different from each other to start with. Also, applying a good heuristic in the fusion process of the model can be quite useful too.

10. Someone in your team has developed a Machine Learning model that has low bias and low variance for a particular dataset. Would you use it for your dataset, which has the same number of variables, but it's inherently different in all other aspects?

A: No, as it's bound to under-perform there. The fact that the two datasets are different (even if they have the same number of variables) makes the usefulness of the model non-transferable. Nevertheless, the model may still be good if a different set of parameters is used, when trained and applied on the new dataset.

Chapter 10

1. Can every statistical model be replaced by a Machine Learning one and still add value to an organization?

A: It's possible in many cases. However, whenever interpretability is a requirement, the Machine Learning model may not be fully equipped to replace a statistical model, even if some Machine Learning models exhibit various levels of interpretability.

2. What constitutes interpretability and why is it important from a business perspective?

A: Interpretability is the characteristic of certain data models, whereby it is possible to interpret their functionality and outputs in an intuitive manner. This is particularly important in business since, oftentimes, it is required that we understand the rationale of certain decisions or predictions and can explain them to others without the use of technical jargon.

3. Would you use a Machine Learning model instead of a statistical to predict the exact demand in energy usage for a factory?

A: Yes, since in a scenario like this predicting the target variable (energy usage) is crucial and any inaccuracies would induce significant costs. So, since performance is of key importance, it would make more sense to employ a Machine Learning model, particularly an advanced one.

4. What kind of a Machine Learning model would you use to tackle a marketing related problem that is linked to the strategy of a company?

A: A transparent model would be best, since the stakeholders of such a project would be interested in the rationale of the predictions made as much as the predictions themselves. Also, in a project like this, performance is not as crucial as in other cases, since there are lots of qualitative factors involved in the decisions involved, so any additional accuracy wouldn't be valued as much.

5. Could you combine a statistical model with a Machine Learning one for better results?

A: It's possible through an ensemble model, although the transparency of the models is bound to diminish significantly, if not disappear altogether.

6. What would you say to a project stakeholder who is convinced that Machine Learning is just a modern way of referring to statistical models?

A: It's best to say nothing in a case like this since it's doubtful you'll change their mind. Instead, you are better off focusing on your work. They may believe the Earth is flat for all you care since their view is not going to change reality in any way.

7. Can you explain to a business person how a Machine Learning model functions, through the use of statistical knowledge only?

A: Unfortunately, not. However, you can say that a Machine Learning model works like a statistical model without the use of distributions or assumptions about feature independence. A Machine Learning model is under the data-driven approach umbrella, and as such, it doesn't rely on assumptions about the data in the way as a statistical model does.

8. What's the value of confidence in a Machine Learning model's predictions when it comes to the application of that model in a business setting?

A: It is quite useful as it can pinpoint predictions less likely to be correct as well as help evaluate the model more accurately (e.g. through the AUC metric in the case of classification).

9. How can you assess the quality of data that you are required to use in a predictive analytics model, for example?

A: Examine how each one of the variables correlates to each other as well as how they correlate to the target variable. Also, calculate the proportion of missing values and outliers in each one of them to ensure that they are worth keeping in the feature set. Optionally, explore also the diversity of the dataset.

10. Are there any statistical models that don't have Machine Learning counterparts, for solving business-related problems?

A: No. For every statistical model there is at least one Machine Learning one, usually yielding better performance, at the expense of transparency.

Chapter 11

1. Why is Machine Learning a constant work in progress?

A: Partly because the field constantly evolves and partly because our understanding of it also evolves, creating new methods over time. Also, the field is relatively new, something that makes it more susceptible to changes, unlike other more established fields, such as Statistics.

2. What's the biggest danger of consuming Machine Learning material these days?

A: Falling prey to opportunists and ignorant people who propagate or even create the hype about the field, usually to serve their own financial interests. This can lead to a warped view of Machine Learning, not to mention disappointment about the field and what knowledge of it can do for you.

3. What's the most important thing when it comes to learning Machine Learning, to avoid getting hustled by opportunists in this domain?

A: In a nutshell, it is acquiring information from reliable sources only (particularly paid ones) as well as ensuring the content creators are knowledgeable about the topic. Also, trying things out in practice is a great way to filter out any information that's either inaccurate or obsolete.

4. What's the deal with the dichotomy between Machine Learning and Statistics when it comes to data models?

A: This is the separation of the main approaches to data models into two categories, those related to Machine Learning methods and those involving statistical methods. However, in practice, this dichotomy is there to illustrate the difference between the two paradigms of data analytics, and it's not representing an exclusive choice of methods. Someone can (and should) use both Machine Learning and Statistics in a data science project.

5. Can Machine Learning and Statistics be combined in a unified framework, when it comes to predictive analytics?

A: Yes, but this would mean that there are significant changes to both Machine Learning and Statistics. A new framework would come about if there is transparency in Machine Learning that makes the two frameworks compatible with each other and the development of a hybrid kind of model a real possibility. If Dr. Wolfram can come up with a mathematical model for the whole universe, perhaps combining Machine Learning and Statistics is not as far-fetched as people think.

6. Why is Julia such a good option for innovating in fields like Machine Learning?

A: Mainly because it's easy to use for prototyping as well as due to its high performance. Also, there is already some existing code you can rely on while developing entirely new methods that run fast is easier than any other high-level language.

Chapter 12

1. Why is Machine Learning considered as the de facto approach to data science in the foreseeable future?

A: Partly because it embodies the data-driven approach in data science, and partly because of the variety of high-performance methods it has under its umbrella (mostly AI-related ones). Also, as it becomes more and more apparent that it can add value to an organization through improved performance, it is expected to gain more popularity cementing its position in data science.

2. What are the most important new methodologies that are under the Machine Learning umbrella? Is this classification justified?

A: Reinforcement learning, semi-supervised learning, and active learning. This classification is largely justified because all of these new methodologies are geared toward AI systems, which are also in the Machine Learning group when they take the role of predictive analytics models.

3. What's the key benefit of the hybrid systems that combine Machine Learning with Bayesian Statistics?

A: The ability to work with limited data, all while yielding high performance and a data-driven approach. In fact, Bayesian Statistics has a lot more in common with Machine Learning than what people think.

4. What is the actually useful AI that's on the rise these days, and how is it different from conventional AI? Also, how is this relevant to the real world (e.g. an organization that wants to utilize this technology in its pipelines)?

A: Interpretable AI, an AI that's not a black box like conventional AI models. Its relevance to the real world lies in the fact that it's less risky and easier to work with.

5. How is Julia expected to evolve in the years to come? How will this affect data scientists?

A: It's expected to grow in different verticals, including packages and a more diverse community of users, along with new material that can make it more accessible to everyone. This is bound to make it a better and more user-friendly tool for many professionals, including data scientists.

6. What's the deal with Gen and why is it relevant to Julia and data science?

A: It is a new programming framework that makes probabilistic and AI-related work easier as it provides an abstraction to the corresponding models, shielding the user from all the heavy math involved. It is relevant to Julia as it was developed on it (it exists as a separate package). As for its relevance to data science, it is bound to be a useful tool for it, particularly for newcomers to the field and for people who want to focus mostly on the bigger picture.

7. What's the single most important thing you can do to ensure that you are in the loop when it comes to Machine Learning and Julia, in the years to come?

A: Continue being involved in the field and maintain a discerning eye, so that you pay attention to the actual trends, not the hype that often surrounds new technologies.

Julia and Other Programming Languages

In this appendix, we'll explore how Julia can be used in tandem with other programming languages. We'll focus on Python, R, and C as they are the ones most commonly used in data analytics these days, though C is there mostly as a deployment language, particularly through its OOP counterparts, C++ and C#. We'll look at bridge packages for connecting Julia with these languages. This involves both running Julia code in those languages as well as running code from those languages in a Julia environment. After all, you are bound to have some code in another language that you don't want to rewrite in Julia, or you may want to utilize a Julia script in another language, taking advantage of the high performance of Julia.

Julia and Python

Call Julia from Python

Linking Julia and Python is quite sensible, considering the latter is the most popular programming language for this sort of work. So, to call Julia from a Python (ver. 3) environment, you need to use the following code (through the PyJulia package which is called "julia" in the package repository of the language):

```
import julia

julia.install("package_name") # install a Julia package

from julia import Base # use the Base package of Julia

Base.sind(30) # sine of an angle of 30 degrees
```

In order for PyJulia to work, you need to have the PyCall package installed as well, something we'll see in the next section. Also, if you use Julia through a Jupyter notebook (i.e. via Ipython), you can use the julia.magic extension:

```
In [1]: %load_ext julia.magic

Initializing Julia runtime. This may take some time...
```

Then, you can run Julia code in the notebook by prefacing it with the %julia metacommand, as for example:

```
In [2]: %julia [1 2 3; 4 5 6] .* 2

Out[2]:

array([[2, 4, 6],

 [8, 10, 12]], dtype=int64)
```

Note that the result is a numpy array. Naturally, this automatic translation of data structures goes both ways:

```
In [3]: arr = [-1, 0, 1]

In [4]: %julia $arr .* 2

Out[4]:

array([-2, 0, 2], dtype=int64)
```

You can learn more about this topic through the documentation webpage of PyJulia: https://bit.ly/2KkQvJw.

Call Python from Julia

Naturally, you can also do the reverse (call a Python script from Julia). To accomplish this, the following code should do the trick (utilizing the PyCall.jl package as well as the NumPy one):

```
using PyCall

p = pyimport("package_name")

p.fun(arguments)
```

If you want to use the "with statement" functionality of Python, you can do so as shown in this example:

```
@pywith pybuiltin("open")("file.txt","w") as f begin

  f.write("hello")

end
```

Should you want to run a series of Python commands and handle variables created in that namespace, you can use the following code and modify it according to what you wish to accomplish:

```
module MyModule

using PyCall

function __init__()

py"""

import numpy as np

def one(x):

return np.sin(x) ** 2 + np.cos(x) ** 2

""" # anything within these triple brackets is interpreted as Python code that is executed
when prefaced by the "py" command

end
```

```
two(x) = py"one"(x) + py"one"(x) # an example of how you can use the abovecustom
```
function as a Julia function

```
end
```

Note that all the data and functions developed in this way are going to be limited in that particular namespace (MyModule) so you won't be able to access them elsewhere in your Julia environment.

Julia and R

Call Julia from R

Since many people from the R world are moving to the Julia one, it makes sense that there is a package in R to link the two environments. Namely, you need to use the following code, utilizing the rjulia package:

```
library(rjulia)

julia_init() # starts a Julia kernel and finds its home folder automatically

julia_eval("julia_expression")
```

Note that this package works with R version 3.1.0 or higher, while the Julia version needs to be 0.5 or higher. Also, since Julia changes quite a bit between releases, it's best to recompile and reinstall the rjulia package every time there is a new version of the language. You can do that using the devtools package, so that you can install the a package directly from github. You can learn more about the rjulia package at its github page: https://bit.ly/3cqxIbw.

Call R from Julia

Although it seems strange, there is also a bridge for calling R scripts from Julia (not that you'll need this option often, but you never know). All this is made possible through the RCall package. The code you need to use is:

```
julia> using RCall

$ # activate R REPL (backspace deactivates it)

R> some_R_command
```

To transfer variables among the Julia and R environments you can use the @rput and @rget macros:

```
julia> x = 123;

julia> @rput x;

R> x

[1] 123

R> y = 5;

julia> @rget y;

julia> println(y)

5.0

julia> a = 4; b = -1

julia> @rput a b

R> a + b

[1] 3.0
```

To execute a series of commands in R, you can use the @R_str macro as follows:

```
julia> y = R"rnorm(30)" # create an array of 30 values following the normaldistribution,
in R, and assign that to variable y
```

```
julia> z = randn(30) # do the same with a julia function

julia> R"t.test($z)" # apply a t-test using the corresponding R function, for the data in z

RObject{VecSxp}

 One Sample t-test

data: `#JL`$x

t = 1.592, df = 9, p-value = 0.1458

alternative hypothesis: true mean is not equal to 0

95 percent confidence interval:

 -0.2116385 1.2172214

sample estimates:

mean of x

0.5027914
```

These examples should be enough to get you started. You can learn more about all this by visiting the corresponding github repository for the RCall package: https://bit.ly/3aouDY9.

Julia and C

Call C from Julia

You can also call C from Julia as several computational geometry functions (such as the N-dimensional convex hull) cannot be found in Julia yet but they are available in C. To utilize a C script you just need to use the following code (based on the ccall package):

```
y = ccall(function_name, output_type, (input))
```

Note that the input part is options since some functions don't require any input in order to run. However, you still need to put the parentheses, e.g. in the case of the *clock* function:

```
y = ccall(:clock, Int32, ())
```

You can acquire data from the C environment through the getenv function as this example for obtaining the path information:

```
path = ccall(:getenv, Cstring, (Cstring,), "SHELL")
```

or this:

```
getenv("SHELL")
```

You can learn more about ccall and other related information for bridging Julia and C through the corresponding documentation webpage: https://bit.ly/2VGXCRx.

Call Julia from C

Although it's unlikely, you may want to call a Julia script from C too. You can manage that using the code that follows (based on the Julia API of that language):

```
jl_init();

jl_eval_string("julia_command");

jl_atexit_hook(0);
```

If you wish to call a particular function with a given argument, you can do that also with the following code:

```
jl_value_t *ret = jl_call1(funcion_name, argument);
```

There is plenty more regarding how you can use Julia and C together, such as handling exceptions, dealing with different variable types, and memory management, all of which you can find in the corresponding documentation page: https://bit.ly/3cqy3ei.

Useful Heuristics Implemented in Julia

In this appendix, we'll look into three useful heuristics that are implemented in Julia, and which can add value to your Machine Learning projects. Namely, we'll look at the *Index of Discernibility*, the *Diversity* heuristic, and the *Performance over Biased Classifier* metric. As a bonus, these heuristics will showcase how you can implement an idea in Julia without the need of relying on the code of someone else, something that, in many cases, may not be even feasible due to the nature of that code. All of these heuristics are my own creations, so they may not be as well-known as other heuristics used in Machine Learning.

The Index of Discernibility heuristic

Overview

This heuristic was developed in four stages. The first two involve my PhD thesis, while the third one on a conference paper I published after my PhD. In each version, the heuristic got faster and more scalable. The latest version is propitiatory for obvious reasons. However, all the other versions of it are open-source. Originally implemented in MATLAB, this heuristic has been translated in other high-level languages, the last one being Julia.

As the name suggests, the Index of Discernibility gauges how discernible the classes of a dataset are. The more discernible, the better a predictor is the input data used. The latter can be a single feature or a set of features. The index of discernibility takes values between 0 and 1, inclusive.

Implementation

A Julia implementation of the Index of Discernibility would be as follows. First of all, we need some auxiliary methods. Namely one for figuring out if a number is within a given range, *within()*, one for getting all kinds of information about the class structure of the dataset, *nc()*, a couple of functions for repeating a number or a vector several times, *rm()*, a basic normalization method using min max, *mmn()*, a sampling method, *sample()*, and a couple of methods for turning a nominal feature or feature set into a set of binary ones, *nom2bin()*. You can find all of these methods in the corresponding .jl script.

Then we have the main variants for the ID metric, one for continuous features, *DID()*, and one for nominal features, *NID()*. Each one of them has two versions, one for the whole feature set and one for an individual feature, both of which are named the same. Note that for the DID metric, the first D stands for distance as it's in reference to the latest publicly available ID algorithm that's based on distances exclusively.

Applications

The Index of Discernibility has various applications. Apart from the obvious one, evaluating the predictive potential of a feature or a set of features, it can be used to gauge how valuable the feature set is before we start applying it to various classification models. This can save us both time and resources as it can aid in refining the "fuel" these models run on. As a result, we can use ID for feature selection, reducing the number of useless features the original dataset has.

Moreover, ID has been shown to help with ensemble models, particularly when combined with a reliability metric. This is especially useful when the members of the ensemble come from different families of models. However, as this is a more advanced topic that requires additional information about reliability metrics, beyond the ones we've discussed in this book, it's best to omit it from this appendix.

Finally, ID can be used for data summarization, which can be used instead of the conventional sampling processes. By selecting the features with the smaller discernibility scores we end up with a more interesting dataset than that of a random sample, since the data points defining the class boundaries will still be there, making the summary dataset more useful in terms of predictive potential.

The Diversity heuristic

Overview

The diversity heuristic is another metric I've developed. It aims to assess how diverse the data in a variable is, be it a continuous one or a categorical one. The diversity heuristic is useful in gauging how diverse the data of a sample is, for example. However, as it's still fairly new, it hasn't been researched thoroughly so it may have other, more interesting, applications.

Diversity in a variable is evaluated by the distances of successive data points in that variable, in relation to the average distance, namely the range of that variable divided by k, where k is the number of data points minus 1. To calculate the exact value of diversity, we need to take the median of these distances and subtract its absolute value from 1 (in other versions of this heuristic, a different metric is used instead of the median). This means that a homogeneous variable would have maximum diversity (1.0), while, in the other extreme, a variable whose majority of data points are gathered in a single point would have the smallest possible diversity (0.0).

Naturally, higher values of diversity are generally better, as they make for a more interesting variable. The same goes for a dataset whose variables all have a high diversity. Although it is possible to define diversity in a multi-dimensional context, it hasn't been done yet due to the distance metrics being unreliable in high dimensionality space. Nevertheless, that's something that could be explored more, given enough research and resources. Feel free to contact me if you are interested in collaborating on such a research project.

Implementation

The latest Julia implementation of the diversity heuristic is as follows. First of all, for a single variable we have the following code:

```
function diversity(x::Array{<:Real, 1}, n::Int64 = length(x))

    sx = sort(x)

    dmax = (sx[end] - sx[1]) / (n - 1)

    if dmax == 0; return 0.0; end

    return 1 - abs(median(diff(sx) .- dmax) / dmax)

end
```

If we wish to calculate the diversity of a whole dataset X, we have a different function for this task:

```
function diversity(X::Array{<:Real, 2})

    n, m = size(X)

    d = Array{Float64}(undef, m)

    for i = 1:m

        d[i] = diversity(X[:,i], n)

    end

    return mean(d), d

end
```

Note that beyond this minimalist version of the diversity heuristic, there is a more complex one that takes into account a different framework for data analytics. However, the latter is enough to fill another book as it redefines almost all existing metrics through a different approach that combines Statistics with AI at a very fundamental level.

Applications

The diversity metric has various potential applications, beyond the evaluation of a sample, in relation to the original dataset that the sample was taken from initially. It can also be used for ensembles as well as for creating synthetic data. As the metric is researched more, additional applications of it can be found.

The Performance over Biased Classifier Metric

Overview

The Performance over Biased Classifier metric is a useful performance metric, applicable in binary classification problems. It is similar to Cohen's Kappa although it covers more performance metrics, plus it offers a more robust evaluation of a classification result, since the biased classifier tends to perform better than the naive classifier (i.e. the classifier that chooses the results entirely by chance, which is the reference point of the Cohen's Kappa metric).

The Performance over Biased Classifier metric takes values between -Inf and 1, with positive values being good and higher values being generally better. In the case of the biased classifier being relatively useless, the PBC metric takes high values. On the other hand, if the biased classifier can offer a decent result (e.g. in the case of the majority class being significantly larger than the minority one), then PBC takes lower values, or even negative ones if the classifier measured fails to deliver.

Implementation

A Julia implementation of the PBC heuristic would be as follows. First of all, we have a couple of auxiliary methods, one for the confusion matrix, *ConfusionMatrix()*, and another one for evaluating different performance metrics, *EvaluateMetric()*, namely the ones mentioned previously. Also, we make use of the *rm()* function that repeats a value several times, significantly faster than the built-in repeat function of Julia. Beyond these methods, we also have the main method of the metric, namely the *pbc()* one as shown here:

```julia
function pbc(yy::Array{<:Any, 1}, y::Array{<:Any, 1}, pm::Int64 = 1)

    N = length(y)

    Q = sort(unique(y))

    q = length(Q)

    nc = Array{Int64}(undef, q)

    for i = 1:q

        nc[i] = sum(y .== Q[i])

    end

    ind = argmax(nc) # majority class

    if (pm == 1) || (pm == 4)

        b = rm(Q[ind], N) # output of a super biased classifier, geared toward the majority
class

    elseif pm == 2

        b = rm(Q[1], N) # output of a super baised classifier, geared

        toward the first class (class 0)

    elseif pm == 3

        b = rm(Q[2], N) # output of a super baised classifier, geared

        toward the second class (class 1)

    end

    zb = EvaluateMetric(ConfusionMatrix(b, y, N), pm, N)

    z = EvaluateMetric(ConfusionMatrix(yy, y, N), pm, N)

    return (z - zb) / (1 - zb)

end
```

Applications

The PBC metric has various potential applications, all geared toward the evaluation of a classifier. This is particularly useful in cases of large class imbalance, resulting in the majority class skewing the performance metric (especially the accuracy rate). Also, it can be used for feature fusion, whereby creating new features through merging the original ones, while optimizing for a performance metric. In this case, the PBC metric can be the fitness function for the optimizations involved.

Index